NextGen Strategies and Tools

NextGen Strategies and Tools: Pioneering Innovations in Agriculture is a systematic, application-oriented examination of next-generation strategies and technologies that drive innovation in agriculture. The book focuses on integrating Artificial Intelligence (AI), Internet of Things (IoT), deep learning, robotics, and smart sensing technologies to address challenges in sustainability, precision farming, and agricultural productivity.

The book presents information on AI-based decision support frameworks, IoT-enabled precision agriculture, smart sensor networks for crop monitoring, and vision-based systems for disease and pest detection. Contributions explore deep learning and computer vision models applied to real-world agricultural scenarios, emphasizing scalability, accuracy, and deployment feasibility in both large-scale and resource-constrained environments. Practical implementations such as Agrobots, Raspberry Pi-based agricultural platforms, and automated fertilizer dispensing systems further demonstrate the applicability of the proposed methods.

In addition to technological advancements, the book examines the sustainability and environmental implications of adopting AI and IoT in agriculture. Key considerations such as efficient resource utilization, climate resilience, and environmentally responsible farming practices are discussed to provide a balanced and holistic perspective. By bridging theoretical foundations with practical deployment, the book supports the development of intelligent agricultural ecosystems.

As part of the *NextGen Agriculture: Novel Concepts and Innovative Strategies* series, this book is intended for postgraduate students, researchers, academicians, and professionals in computer science, AI, data science, electronics and communication engineering, agricultural engineering, and smart farming technologies. It also serves as a reference for policymakers and practitioners seeking to advance sustainable and intelligent agricultural systems.

NextGen Agriculture: Novel Concepts and Innovative Strategies
Chittaranjan Kole

Series Editor

Allele Mining for Genomic Designing of Fruit Crops
Chittaranjan Kole, Kenta Shirasawa, and Anil Kumar Singh

Allele Mining for Genomic Designing of Cereal Crops
Chittaranjan Kole, Sharat Kumar Pradhan, and Vijay K Tiwari

Allele Mining for Genomic Designing of Vegetable Crops
Chittaranjan Kole, Tusar Kanti Behera, and Prashant Kaushik

Allele Mining for Genomic Designing of Oilseed Crops
Chittaranjan Kole, Manish Kumar Pandey, and Naveen Puppala

Sustainable Urban Agriculture: New Frontiers
Kheir Al-Kodmany, Madhav Govind, Sharmin Khan, and Chittaranjan Kole

Allele Mining for Genomic Designing of Grain Legume Crops
Chittaranjan Kole, C. Bharadwaj, and Abhimanyu Sarkar

The Vertical Farm: Scientific Advances and Technological Developments
Kheir Al-Kodmany, Andrew Keong Ng, Abel Tablada, and Chittaranjan Kole

Next Generation Food Crops for Human Health
Edited by Sangam Dwivedi, Rodomiro Ortiz, and Chittaranjan Kole

Smart Agriculture: Concepts, Strategies, and Case Studies
Edited by Verónica Saiz-Rubio, Francisco Rovira-Más, and Chittaranjan Kole

Tailoring the Tomato Genome for NextGen Agriculture
Edited by Dilip R. Panthee and Chittaranjan Kole

NextGen Strategies and Tools: Pioneering Innovations in Agriculture
Edited by Dhakshayani J, Shashidhar R, Vinayakumar Ravi, Roopashree S, and Deepti Gupta

For more information, please visit our website: https://www.routledge.com/Nextgen-Agriculture/book-series/NGA

NextGen Strategies and Tools

Pioneering Innovations in Agriculture

Edited by
Dhakshayani J, Shashidhar R, Vinayakumar Ravi,
Roopashree S, and Deepti Gupta

CRC Press
Taylor & Francis Group
Boca Raton London New York

CRC Press is an imprint of the
Taylor & Francis Group, an **informa** business

Designed cover image: Shutterstock

First edition published 2027
by CRC Press
2385 NW Executive Center Drive, Suite 320, Boca Raton FL 33431

and by CRC Press
4 Park Square, Milton Park, Abingdon, Oxon, OX14 4RN

CRC Press is an imprint of Taylor & Francis Group, LLC

ISBN: 9781041008361 (hbk)
ISBN: 9781041011774 (pbk)
ISBN: 9781003613510 (ebk)

DOI: 10.1201/9781003613510

Typeset in Times
by Newgen Publishing UK

Dr. Norman Ernest Borlaug
The Father of the Green Revolution,
Nobel Peace Prize winner and the Founder of the World Food Prize

Who inspired my research and academic works with his generous appreciation and advice.

Preface to the Series

Agriculture is going through a phase of transformation now. It has both challenges and opportunities. One of the major challenges it is facing is obviously food security. A population of about 9.3 billion projected by 2050 necessitates an increase of crop production by 60–70%. Another 10–15% increase would be needed to counter the reduction of crop production due to global warming and climate change. Nutrition security is another challenge with more than 800 million people suffering from malnutrition. Depletion of fossil fuel and increasing dependence on biofuel from crop residues is another challenge. Loss of biodiversity and land degradation also add to the list of challenges. The cultivable area is also decreasing fast due to housing, transport, and industrialization. Above all, climate change and global warming have a huge impact on agriculture, health, and environment.

Several novel concepts and strategies are emerging to address the above challenges. For example, the practice of precision farming will lead to precise and timely use of inputs resulting in higher yield and better quality of produce on the one hand and reduction of loss of inputs on the other. Similarly, urban agriculture and vertical agriculture will add a lot of cultivable areas in the future. Protected farming will facilitate growing crops under controlled conditions, leading to reduction in biotic and abiotic stresses and precise utilization of inputs.

Agriculture requires maintenance of the philosophy farmers have been following since inception over thousands of years which are being sold today in new packages with sophisticatedly coined labels such as traditional agriculture, natural agriculture, sustainable agriculture, conservation agriculture, integrated agriculture, regenerative agriculture, organic agriculture, and so forth.

In the changing scenarios, next-generation agriculture will need designed crop varieties and breeds produced through designed practices. This book series aims at deliberations on several novel concepts and innovative strategies developed for next-generation agriculture and will hopefully benefit agricultural education, research, and outreach.

Kolkata Chittaranjan Kole
January 10, 2024

Special Acknowledgement to
Phullara Kole
For her outstanding assistance in editing this book series

Contents

About the Editors

Dhakshayani J is currently working as an assistant professor in the Department of Computer Science and Engineering at the Indian Institute of Information Technology Kottayam, Kerala. She completed her Ph.D. in Computer Science and Engineering from the National Institute of Technology, Puducherry and obtained her Master's degree from Pondicherry University, Pondicherry. Her research domain focuses on precision agriculture, with a strong emphasis on applying image processing and computer vision techniques to real-world agricultural challenges. Her work spans crop disease detection using visual analytics, automated plant trait analysis, and high-throughput phenotyping through multimodal data fusion. She is currently leading an ongoing ANRF-sponsored research project titled "AI-Driven Multimodal Vision System for Rubber Tapping Optimization Using RGB, Thermal, and 3D Imaging in Kerala Plantations," which focuses on integrating multiple visual modalities to enhance tapping efficiency, plant health assessment, and decision-making in rubber plantations. In addition to her research activities, Dr. Dhakshayani J serves as an active reviewer for several reputed international journals, including Scientific Reports, Wiley, Hindawi, and several Springer and IEEE international conferences.

Deepti Gupta is an assistant professor at Texas A&M University-Central Texas. After receiving her PhD, she joined Goldman Sachs as a cloud security architect. She also worked as a faculty member in the Department of Computer Science at Huston-Tillotson University, Austin. She received her Ph.D. degree in Computer Science from the University of Texas at San Antonio (UTSA) and also received her M.S. degree in Computer Science from UTSA. She has worked as an adjunct faculty in the Department of Computer Science at St. Edward University, Austin. Dr. Gupta's research interests lie in the areas of security and privacy in the Internet of Things (IoT), leveraging cloud and edge computing. Her research interests also include the application of AI and Machine Learning to secure IoT and CPS infrastructures in various application domains, such as smart healthcare, wearable IoT, and smart home. She is also interested in designing federated learning algorithms to deal with non-IID data using game theory. She also developed novel anomaly detection models and fine-grained access control models to develop secure infrastructure for IoT. She has several conference and journal publications and also continually serves as an expert reviewer for various journals and technical program committees for several conferences and workshops. Dr. Gupta has received the National Science Foundation (NSF) award. She is an active team member of IEEE ComSoc Young Professionals, AnitaB.org, WiCyS, and also co-chair of the N2Women fellowship.

Vinayakumar Ravi (https://vinayakumarr.github.io/) is an assistant research professor at the Center for Artificial Intelligence, Prince Mohammad Bin Fahd University, Khobar, Saudi Arabia. His previous position was a postdoctoral research fellow in developing and implementing novel computational and machine learning algorithms and applications for big data integration and data mining with Cincinnati Children's Hospital Medical Center, Cincinnati, OH, USA. His current research interests include applications of data mining, Artificial Intelligence, machine learning (including deep learning) for biomedical informatics, cyber security, image processing, and natural language processing. He has more than 100 research publications in reputed IEEE conferences, IEEE Transactions, and Journals. His publications include prestigious conferences in the area of cyber security, like IEEE S&P and IEEE Infocom. Dr. Ravi received a full scholarship to attend Machine Learning Summer School (MLSS) 2019, London. He has organized a shared task on detecting malicious domain names (DMD 2018) as part of SSCC'18 and ICACCI'18. He received the Chancellor's Research Excellence Award in AIRA 2021, and his name was included in the World's Top 2% Scientists by Stanford University published in PLoS Biology.

Roopashree S., currently an associate professor in the Department of Computer Science and Engineering, B.M.S. College of Engineering, Bull Temple Road, Bengaluru, KA, India, is a highly motivated and accomplished professional with over 16 years of experience across academia and industry. She has achieved significant milestones in her academic journey. Dr Roopashree is a dedicated researcher in the fields of artificial intelligence, computer vision, machine learning, generative AI, large language models, small language models, and image processing. She has an impressive track record, including the Indian utility patent grant and publication of three Indian patents and the grant of a German utility patent and two Australian design patents. As a senior member of IEEE and a lifetime member of the Computer Society of India, Dr Roopashree has made significant contributions to both research and academia. She has published 35+ research articles indexed in Web of Science, SCIE, and Scopus Journals/conferences. She is an active reviewer for leading SCIE journals. Her achievements include working on a funded multi-agent framework project on Indian medicinal plants, serving as an R&D consultant to start-ups, receiving sponsorship from the Karnataka State Council for Science and Technology (KSCST) for three projects, and securing first place in the IEEE Humanitarian Projects Proposal competition in 2020.

Shashidhar R received his Bachelor of Engineering and Master of Technology degrees from Visvesvaraya Technological University (VTU), Belagavi, and his Ph.D. from JSS Science and Technology University (JSS STU), Mysuru. He is currently working as an associate professor in the Department of Electronics and Communication Engineering and serves as the Convener of the Institute Innovation Council at JSS Science and Technology University, Mysuru. He has published over 60 papers in Scopus- and SCI-indexed journals and has edited three Scopus-indexed books. He is an editorial board member of Scientific Reports (Nature Publishing Group) and Discover Imaging (Springer Nature). He has filed nine patents, of which seven have been granted and two are under final examination. He received the Best Paper Award for the paper titled "Music Emotion Recognition Using Convolutional Neural Network for Regional Languages" at AIKIIE-2023, Ballari. He was awarded Best Paper Presenter for the paper "Optimizing EMG Signal Analysis with Advanced Machine Learning Techniques" at ICSTEMSD-2025. He also received the Best Paper Award for "Speech Emotion Recognition Based on Speaker and Gender Dependences in Cross-Linguistic" at CCICT-2025. He received the Protsahan: Recognition of Research Publications Award from the IEEE Communications Society, Bangalore Section, during January 2020–September 2021 and October 2021–September 2022. He is an Executive Committee Member of the IEEE Bangalore Section and serves as the Secretary of the IEEE Mysuru Subsection. He has served in various roles in IEEE international conferences, including Technical Program Committee Chair at IEEE MysuruCon 2023 and ICETEG-2025, and Track Chair at IEEE MysuruCon 2022 and INDISCON 2023. He has over 1,000 Google Scholar citations, with an h-index of 17 and an i10-index of 27. His research interests include speech processing, biomedical signal and image processing, embedded systems, artificial intelligence, machine learning, deep learning, and quantum machine learning.

Contributors

Aarthy C
Department of CSE(AIML), SRM TRP Engineering College, Tiruchirappalli, Tamil Nadu, India

Akilandeeswari M
Department of Information Technology
PPG Institute of Technology, Coimbatore, Tamil Nadu, India

Balakrishnan D
Department of Computer Science and Engineering (Artificial Intelligence and Machine Learning), K.S. Rangasamy College of Technology, Namakkal, India

Bharath B C
Information Science and Engineering Department, Dayananda Sagar College of Engineering, Bangalore, India

Guda Bhargavi
Department of Agronomy, SR University, Warangal, India

Lakshmi Bhaskar
Department of ECE, BNM Institute of Technology, Bangalore, India

Charithra C M
Department of Business Administration, BNM Institute of Technology, Bangalore, India

Chayashree G
Vidyavardhaka College of Engineering, Mysore, Karnataka, India

Chethana H.T.
Department of Computer Science and Engineering, Vidyavardhaka College of Engineering, Mysuru, India

Deebika R
Computer Science and Engineering, K.S.R. College of Engineering, Tiruchengode, India

Shridevi A Desai
Department of Machine Learning, B.M.S. College of Engineering, Bull Temple Road, Bengaluru, KA, India

Deepti Gupta
Department of Computer Information Systems, Texas A&M University–Central Texas, Texas, USA

Hemavathi
Ramaiah Institute of Technology, Bengaluru, Karnataka, India

Varun Jayadeva
University of Birmingham, England, United Kingdom

Jagadamba G
M S Ramaiah University of Applied Sciences, Bengaluru, Karnataka, India

Archana Kasula
Department of Soil Science and Agricultural Chemistry, Professor Jayashankar Telangana Agricultural University, Hyderabad, India

Kasula Vamshi Krishna
Department of Agronomy, SoAS, Malla Reddy University, Hyderabad, India

Hong Lin
University of Houston–Downtown, USA

Madhura J
Information Science and Engineering Department, Dayananda Sagar College of Engineering, Bangalore, India

Neroju Manasa
Department of Agronomy, SR University, Warangal, India

Manjunath A S
Department of Computer Applications, JSS Science and Technology University, Mysuru, India

Narmada C
Department of Computer Science and Engineering, SNS College of Technology, India

Vani Pradhayi G L K
School of Computer Science and Engineering, Vellore Institute of Technology, Chennai, India

Parvathi M.
Artificial Intelligence and Data Science, Nandha Engineering College, Erode, India

Prathyusha P
Department of Business Administration, BNM Institute of Technology, Bangalore, India

Ravi P
AP/CSE(AIML)
KPR Institute of Engineering and Technology, Coimbatore, Tamil Nadu, India

Vigrahala Rajeshri
School of Computer Science and Engineering, Vellore Institute of Technology, Chennai, India

Rachana M S
Department of Computer Science and Engineering, B.M.S. College of Engineering, Bull Temple Road, Bengaluru, KA, India

Roopashree S
Department of Computer Science and Engineering, B.M.S. College of Engineering, Bull Temple Road, Bengaluru , KA, India

Shashank M P
Department of Electronics and Communication Engineering, JSS Science and Technology University, Mysuru, India

Shashidhar R
Department of Electronics and Communication Engineering, JSS Science and Technology University, Mysuru, India

Shalini M S
Department of Electronics and Communication Engineering, JSS Science and Technology University, Mysuru, India

Sharmiladevi S
School of Computer Science and Engineering, Vellore Institute of Technology, Chennai, India

Rakshitha D Shastry
Department of Computer Science and Engineering, Vidyavardhaka College of Engineering, Mysuru, India

Rajeshwari J
Information Science and Engineering Department, Dayananda Sagar College of Engineering, Bangalore, India

Renuga Devi B
Department of CSE(AIML), SRM TRP Engineering College, Tiruchirappalli, Tamil Nadu, India

Richa Sinha
Department of Computer Science and Engineering, Vidyavardhaka College of Engineering, Mysuru, India

Shalini K B
Information Science and Engineering Department, Dayananda Sagar College of Engineering, Bangalore, India

Sumathi A
Department of ECE, BNM Institute of Technology, Bangalore, India

Sunitha S V
Department of ECE, BNM Institute of Technology
Bangalore, India

Vasuki R
Department of Computer Science and Engineering, Dhanalakshmi Srinivasan University, India

1 An Integrated Approach for Sustainable Agriculture Development Using NextGen Artificial Intelligence Solutions

Prathyusha P and Charithra C M

1.1 INTRODUCTION

A significant portion of India's population resides in villages and is engaged in farming. The agriculture industry brings the connection between the markets of urban and rural areas. The produce from the agriculture industry is supplied to the markets in the urban areas, thus creating an increase in an organized transportation model, storage facility, and a well-connected distribution network. This makes it important to prioritize rural development for sustainable growth (Kang et al., 2014; Sahoo et al., 2012).

Innovating existing practices and adopting sustainable agriculture through analysis of critical factors will help achieve better rural livelihoods along with a green economy and food security. Global agricultural communities encounter several problems including changes in weather dynamics, soil mechanics, water shortage, pest management, and an increase in nutrient food requirements (Bouteska et al., 2024; Intergovernmental Panel on Climate Change (IPCC). 2014). Traditional farming practices produce favorable results, yet agricultural professionals such as farmers and marketing teams struggle with a few challenges. We need sustainable agricultural development methods that maximize yield through efficient resource management while protecting the environment from future hazards caused by global warming and extreme weather events (AlZubi, 2023).

Exploration of new concepts and solutions requires the collection of the most recent innovations in NextGen AI Solutions for all data types, including spatial and aspatial data of both relational and unstructured data formats. These solutions enable us to combine automation with predictive analytics and data-driven decision-making through technological disruption. The combination of machine learning, computer vision, robots, and the Internet of Things (IoT) enables AI-driven technologies to power smart farming with near real-time monitoring and autonomous farm management (Gandhi et al., 2020). The latest technology adaptation benefits to improve crop yields while establishing a future of sustainable farming practices that safeguard both food security and environmental equilibrium (Pandya et al., 2019).

1.1.1 Role of Artificial Intelligence in Sustainable Agriculture (SA)

Artificial Intelligence (AI) for sustainable agriculture consists of the latest technology adaptation and its usage for smart farming. This improves productivity, sustainability, and efficiency by meeting future demands. AI in agriculture uses industry-proven ML algorithms, continuous data analysis, and IoT devices to gather and interpret critical information. The simplified user-friendly

DOI: 10.1201/9781003613510-1

AI solutions will help farmers at different skill levels to arrive at quick and appropriate analysis to choose the right action, leading to early detection improvement and correction factors (Ma, 2022).

The principle of sustainable agriculture is it uses NextGen AI solutions to analyze a large amount of spatial data with minute objects sizing up to one foot (Wihardi, 2020) improving the accuracy of the outcome, Internet of Things (IoT) sensors, Aerial Photography and surveillance for cloud seeding, and drones to collect the temporal data. The information is vital for checking the daily status of fields, which would enable accurate interpretation.

1.1.2 An Integrated Approach for Sustainable development in Agriculture with NextGen AI solutions

The integrated approach of NextGen AI solutions consists of the following major steps:

1.1.2.1 Usage of Data Models in SA

The predictive models are used to analyze the historic data and consider the present weather conditions and soil fertility in order to forecast the crop outcome with greater accuracy as shown in Figure 1.1. This allows the farmers to plan better and reduce risks. By understanding weather trends and soil conditions, AI can suggest the best planting times and crop varieties for maximum yield.

Machine learning is classified into three types based on the algorithms:

1.1.2.1.1 Supervised Machine Learning (SML)

Supervised machine learning models are trained with "well-labelled" data points. This implies that the data at the input are correctly associated with the output, using which the machine now experienced to make more accurate forecasts (Feng, 2019). In supervised learning, the "labelled data" acts as a guide, similar to how a teacher helps a student learn. This process involves feeding the input and the corresponding output into the ML model, with an objective to help the SML model identify a function that maps the input variable (x) correctly to the output (y). The process is represented in Figure 1.2.

Types of Supervised Learning

Classification:

The classification technique is based on relative comparison and deriving the results based on previously trained data sets. For example, the financial matters and behavior of the individuals can be

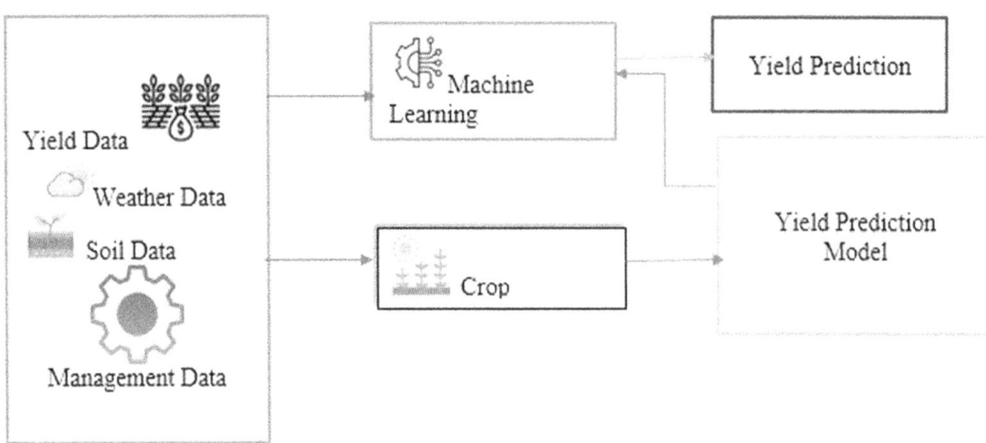

FIGURE 1.1 ML-based prediction model.

TABLE 1.1
ML Classification Algorithm

Classification Algorithm	Cases
Artificial neural networks (ANN)	It is a complex classification algorithm, commonly used for satellite data recognition, Natural Language Processing (NLP), etc. This is appropriate in image analysis and is vital to monitor and predict crop yields
Random forest	This classifier integrates one or more decision trees and arrives at a more accurate prediction model
k-Nearest neighbor (kNN)	It is a non-parametric algorithm that classifies based on the nearest values from the trained data sets
Decision tree	This is the easiest classifier for obtaining insights and drawing interpretation and can be recommended to apply to new data
Logistic regression	A stochastic algorithm is often referred to get the relationships among variables and predict the potential outcomes based on trained data sets

compared, and the banks can guess the future characteristics, including financial well-being, and conclude better possibilities for the individuals to repay the loan amounts.

In the example given, the financial behavior is referred to as features and the variables of payment history are mapped as targets. Classification is expressed in binary mode either 1 or 0.

The algorithms are built on the concept of guessing, and it is also called as binary classification as it has only two options (yes or no, true or false). For more than 1 option, it is referred as multiclass classification.

ML classification algorithm techniques along with the cases used is presented as follows:

Regression

Regression is a fundamental technique in supervised machine learning, that is based on understanding the "data with relationship analysis". The significant application is seen in the analysis of company sales, employee salary, health predictions, etc. This is a technique that is different from the task of classification, where the objective is to classify the appropriate category for any new data. Whereas regression analysis identifies significant connections between the various factors and the expected outcome we wish to predict. Regression analysis uses multiple machine learning algorithms which helps is prediction that includes:

- **Linear Regression**: This is a prediction technique that makes use of all the known data variables to estimate the value of an unknown data point. The linear relationship between the variables is represented by

$$Y = aX + b$$

where,
Y – dependent variable (in this case yield, weather, soil and management data)
a – slope (factors like water, pest removal etc)
X – independent variable

- **Decision Tree Regression**: This is a technique used for both classification and regression problems consisting of categorical and continuous dependent variables. In this algorithm, the data population gets split into two or more homogeneous sets with the most significant attributes/independent variables making into distinct groups.

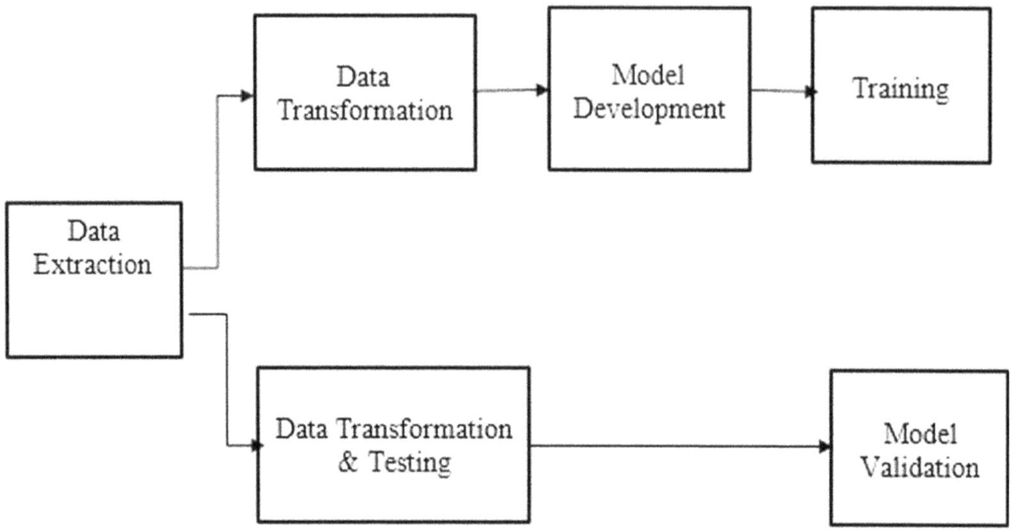

FIGURE 1.2 Steps involved in supervised machine learning.

i. **Data Extraction**
The process begins with identifying the problem to be solved. The agricultural data characteristics like the impact of weather, changes in the soil fertility, etc, are to be studied. Upon identifying the data category, all relevant attributes required are presented. It is very crucial that the data after sorting suits the requirements to ensure the prediction model is accurate.

ii. **Data Transformation**
Smart framing is used to align the data extraction to various units of data collection. Each of these are defined by the logical data model to ensure consistency in data and for a relevant and reliable outcome. Once the image and video data extraction is set up, the most important hurdle is to eliminate anomalies if any, remove the false positives and noise. The appropriate data filter will be applied on this to split the data into two parts. 80% of the data will be used for training and another 20% to test based on the Pareto principle. Verification of data and then validation is essential to evaluate the trained inputs.

iii. **Development of the Prediction Model**
Prediction models like ANN or KNN are used for data in image format while for data in continuous format like soil fertility, weather, or pest control are predicted with the help of linear and logistic regression.

iv. **Training the Prediction Model**
The best fit algorithms are used to improve the accuracy of the prediction. This is a continuous process where the model gets trained. The model is adjusted by repetitive training and using the knowledge gained. The complex settings of the model are managed by setting the validation rules, which in turn improves the correctness of the outcome.

v. **Model Validation**
This is a process of examining the reliability of the model by testing the consistency and accuracy of the prediction it makes. The effectiveness of the models is evaluated on the data set it hasn't seen before. This becomes an important step because it helps identify the parameters and problems, if any, before implementation.

1.1.2.1.2 Unsupervised Machine Learning

This technique uses machine learning algorithms to group the data sets. Without human intervention, they are capable of picking the hidden patterns. This feature makes it ideal for usage

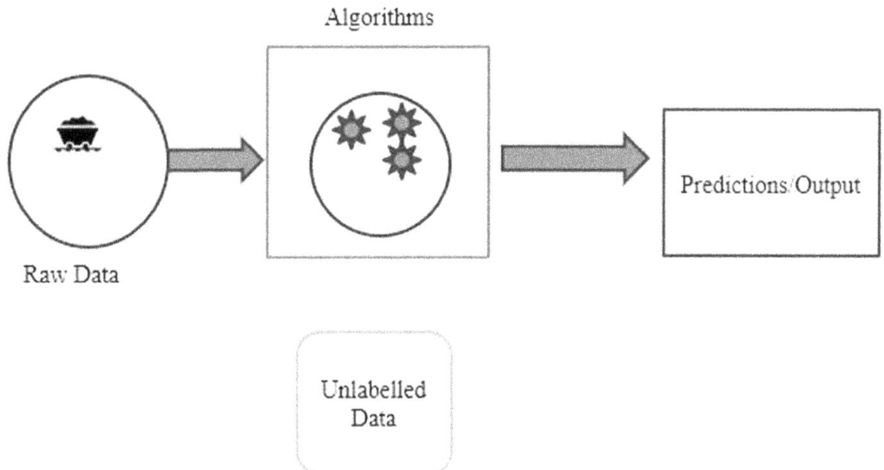

FIGURE 1.3 Unsupervised learning process.

in exploratory data analysis. The major tasks performed by it are explained in Figure 1.3 as follows:

Algorithms Used in Unsupervised Learning:

Clustering: This is a mining technique which clusters or groups the data sets based on the differences or similarity. This is also an unsupervised machine learning algorithm which helps in analyzing and understanding unlabeled data sets.

Important clustering algorithms used in Industry are as follows:

- K-Means
  K-Means clustering algorithm is the most common technique used for clustering. The centroid-based algorithm is used to define the clusters of unlabeled data. This is an iterative algorithm that creates nonoverlapping subgroups meaning each instance belongs to an exclusive cluster.
- Hierarchical Clustering
  This machine-learning method segregates the data into tree-like groups called clusters. Hierarchical clustering can be performed with both bottom-up and Top-down observation merge and split functions.
- Fuzzy-C-Means (FCM):
  FCM method uses the application of the principles of Fuzzy logic for data clustering.

Dimensionality Reduction (DR): It is an unsupervised machine learning technique that is used to reduce the dimensions or input data variables in the dataset to make the process smoother. It can also be understood as techniques used in the classification and regression of datasets to create effective predictive models.

There are many techniques in the dimensionality reduction (DR) method. The popular ones are:

1. Linear discriminant analysis (part of supervised machine learning)
2. Principal component analysis (part of unsupervised machine learning)

Association: An association rule is used to find relationships between variables in large databases. It identifies data groups with frequency in the dataset. Association rules can enhance marketing strategies. For example, people who buy item X (like car) are also likely to purchase item Y (may be petrol or diesel).

When to use Unsupervised Machine Learning (USML) in NextGen AI SA solution:
Below are the best cases where USML can be implemented:

- USML helps uncover hidden insights from a dataset. Since it does not have labelled datasets, it learns from its own experience like a human. And that's what makes it a real AI.
- In reality, we may not always have data sets that are labelled. Hence, this technique becomes an important application for a new data set.

Challenge: Here, the magnitude of accuracy is not as high as in supervised machine learning cases due to incomplete information on the input data labels and accurate expected output.

1.1.2.1.3 Reinforcement Learning

This is a subset of machine learning. Here, the agent learns decision-making by interacting with the environment through trial and error. Hence, this does not need feeding data or programming for every possible scenario as it adapts its decision-making capacity dynamically. It also has the capacity to learn continuously and improve based on the feedback. Q-learning, Deep Q Network (DQN), Policy Gradients, and Actor-Critic methods are a few popular RL algorithms to name. The major difference among these methods lies in how they represent and learn policies and value functions. For example, Q-Learning is trained with optimal policy by learning a value function that gives the maximum assignments to the state-action pair. DQN extends Q-Learning to higher dimensions using deep neural networks to represent the action-value function. Policy gradients, alternatively, directly optimize the policy without the need for a value function (Mahibha, 2023).

1.1.2.2 Pest Identification and Control Measures

Automated image recognition can identify and segregate pests. This helps in formulating a targeted pest control mechanism. Artificial intelligence is used to recognize the species of pest from the images captured on the site which warns the farms early thus reducing the extent of damage as shown in Figure 1.4. This mechanism can be used to dynamics of the pest category and effective management strategies (Indu, 2022).

1.1.2.3 Precision Farming

Precision farming works on the principle of using and maximizing the benefits from various farming practices with the objective of causing minimum impact on the environment while maintaining productivity (Idoje, 2021).

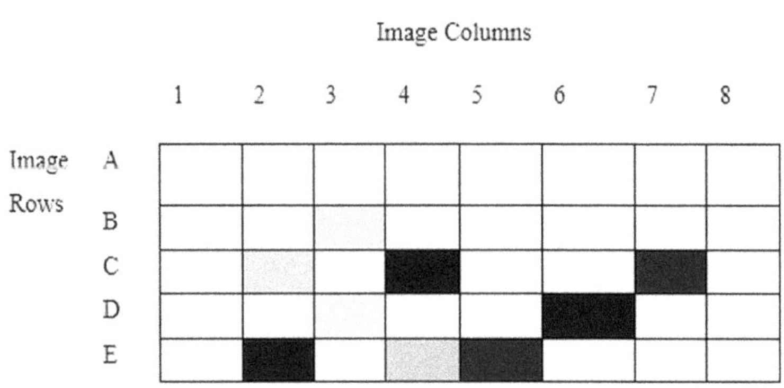

FIGURE 1.4 Integrating satellite, drone, and aerial photography images to detect the pest species.

Precision farming, or precision agriculture, is the application of high technologies to maximize field-level management for crop farming. The process incorporates data-based methods to improve productivity, efficiency in resource use, and sustainability. The main steps (Figure 1.5) of the precision farming process are as follows:

Data Collection

The process starts with the methodical gathering of information from remote sites. The sources are remote sensing (satellites, drones), field sensors (soil moisture, temperature, pH), GPS devices, and meteorological stations. Farmers could also compile historical crop and harvest information.

Data Analysis and Interpretation

Collected data is interpreted and analyzed with Geographic Information Systems (GIS), artificial intelligence (AI), or data analytics software to recognize spatial and temporal variability in the field. This assists in the interpretation of field conditions, disease epiphenomena, soil health, and crop demands.

Decision-Making and Planning

Decision-making for site-specific management is made on the basis of data analysis. These can range from establishing the type and amount of fertilizers, ideal scheduling of irrigation, pest and disease management, and planting strategies. Variable rate technology is usually utilized for accurate input application.

Implementation of Field Operations

The intended interventions are applied through automated or semi-automated machinery that has GPS and IoT-based control systems. This provides precision seeding, fertilizing, irrigation, and pesticide application, with reduced waste and increased efficiency (Kang, 2014).

Monitoring and Evaluation

Following implementation, monitoring crop performance and field conditions is constantly carried out to evaluate the impact of interventions. This stage aids in adjusting needed actions and enhances future decision-making loops.

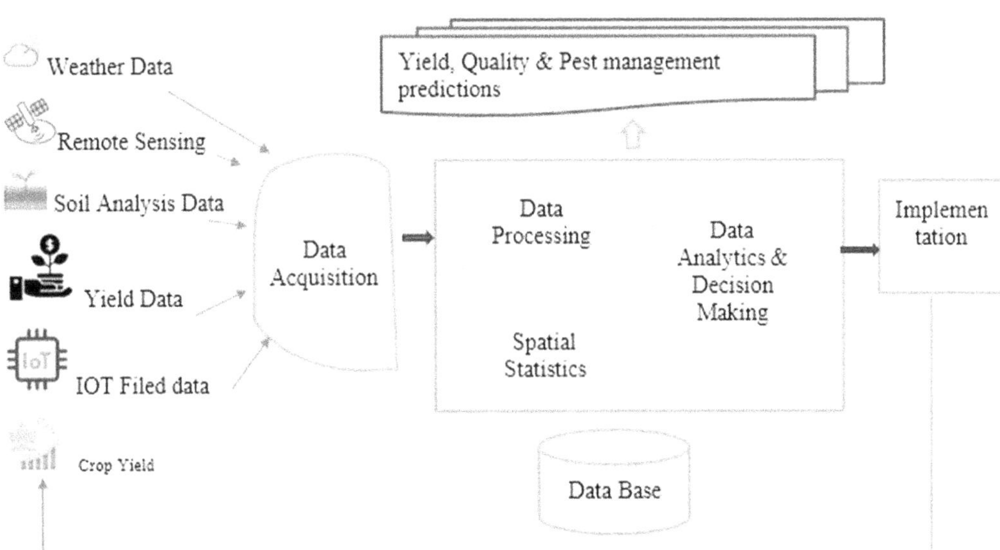

FIGURE 1.5 Data analytics-based precision farming.

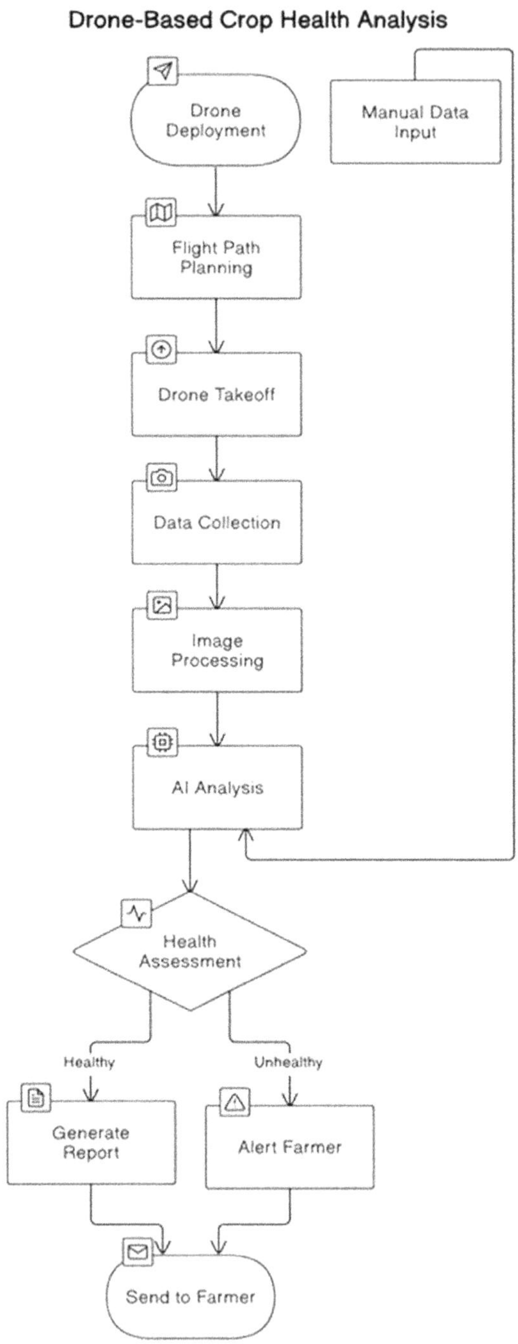

FIGURE 1.6 Unmanned aerial surveillance.

Feedback and Optimization

Lastly, information for the ongoing crop season is saved and applied to calibrate models and strategies for future seasons. On going learning and optimization are part of improved precision farming results in the longer term.

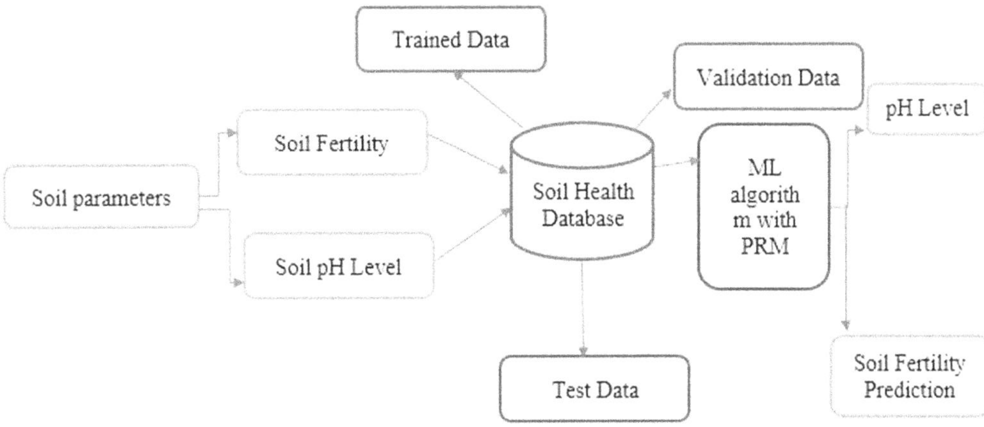

FIGURE 1.7 Soil quality prediction model.

1.1.2.4 Unmanned Aerial Surveillance

The unmanned aerial vehicles supervise the health of the crops aerially, which is crucial information for the farmers to control damage. It helps in large fields surveillance, providing information on nutrient and water deficiency (Qazi, 2022). NextGen AI SA solutions suggest the usage of drones usage to improve the response rate to the problems enabling reduction in damage and hence improving the productivity. The process is explained in Figure 1.6 as follows.

1.1.2.5 Soil Quality Prediction with ML

The ML algorithms are used to test the samples of the soil to determine the optimum nutrient levels and suggest measures to improve the fertility of the soil (Adamchuk, 2004). AI is capable of predicting the right crop considering the seasonal challenges (Kumar, 2022). This leads to better crop yields and sustainable usage of the natural resources. The process is showed in Figure 1.7.

1.1.2.6 Smart Irrigation Systems

Water usage by the crop can be optimized by using AI algorithms. These algorithms will analyze the level of moisture in the soil, forecast the weather, and the exact requirement of water to the crop (Figure 1.8). ML programs like multivariate-linear regression, logistic regression (LR), support vector machine, decision tree (DT), random forest (RF), K-means, Bayesian inference (BI), and artificial neural network (ANN) are widely used in determining the soil conditions hence optimizing the water requirement (Jalal, 2021).

1.1.2.7 NextGen AI SA Sorting Robots

The final product inspection is automated to ensure accurate grinding thus increasing the profits. This allows the managers to deploy manpower based to different tasks. These work 24 × 7 and gives the resource efficiency (Arlitsch, 2017). It incorporates cameras that analyze products on a rolling table and applies AI models to validate their visual traits (Gandhi, 2020). Based on its findings, the AI technology triggers robotic sorting arms to remove unacceptable products at rates between 80 and 100 picks per minute. This entire process is represented in Figure 1.9.

1.1.2.8 Predicting the Market Demand

Artificial intelligence algorithms predict the demand, which helps in organizing the entire supply chain. Thus decreasing the wastage of food and ensuring fresh supply reaches the market (Figure 1.10).

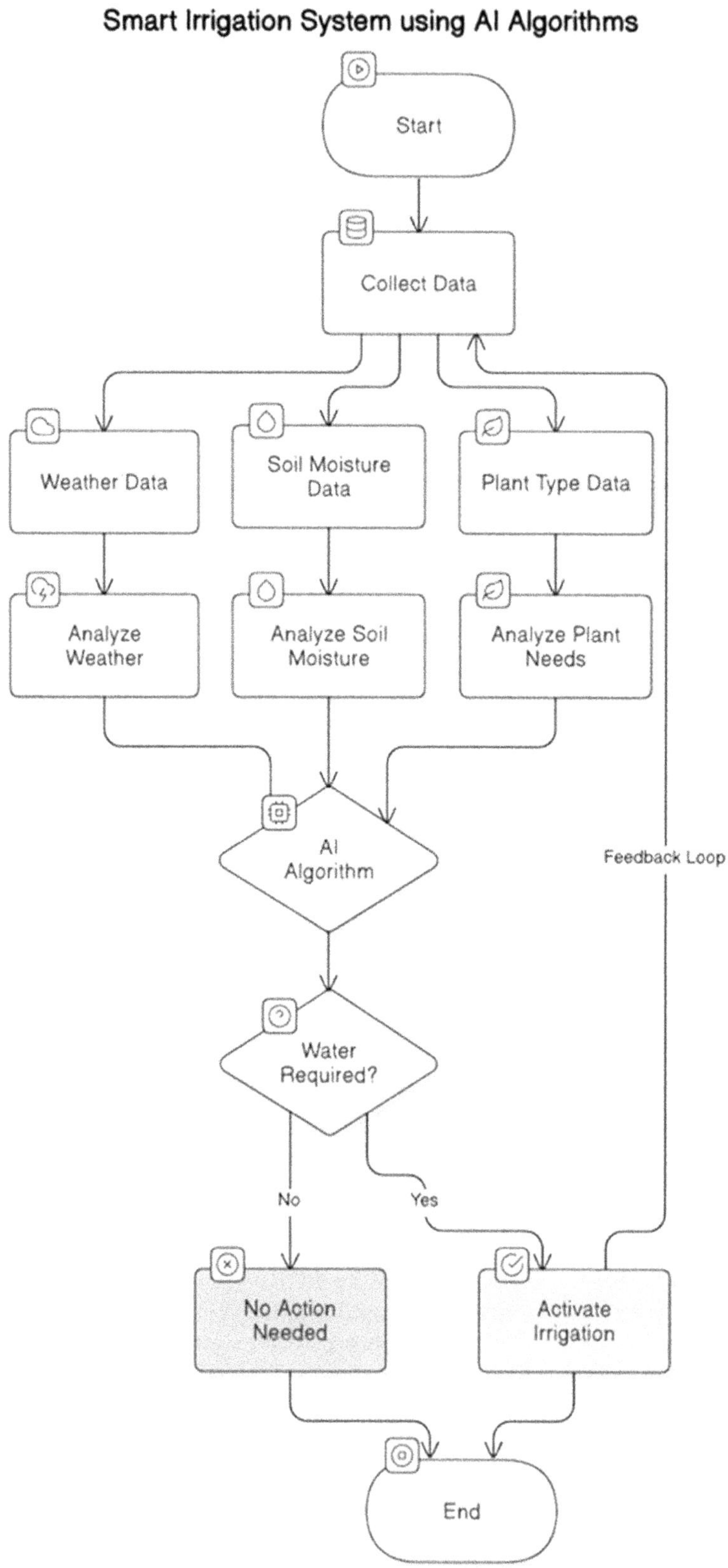

FIGURE 1.8 Smart irrigation systems.

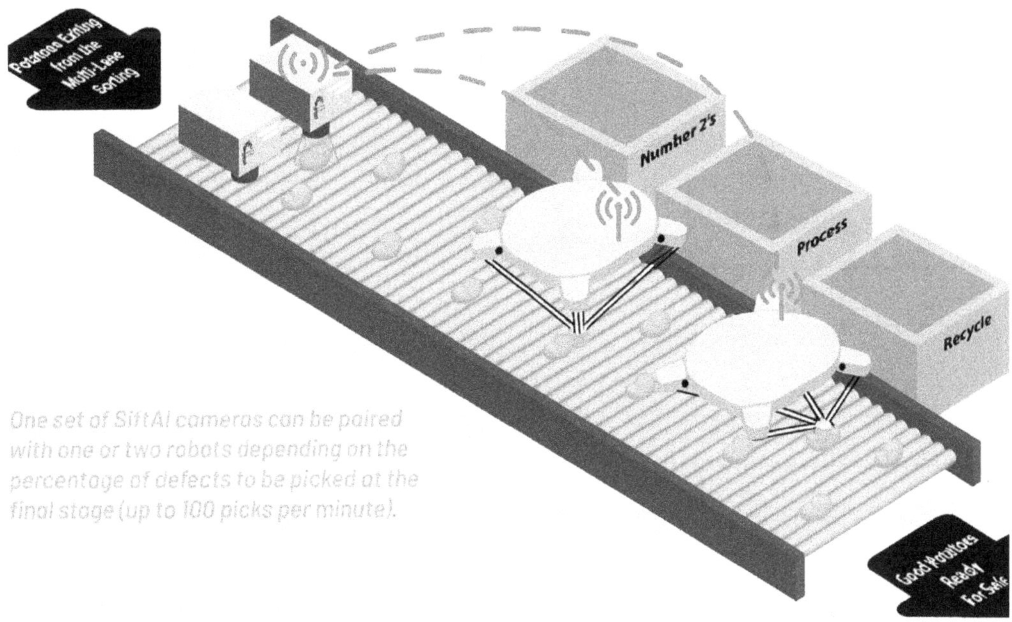

FIGURE 1.9 Sorting robots in NextGenAI SA solution.

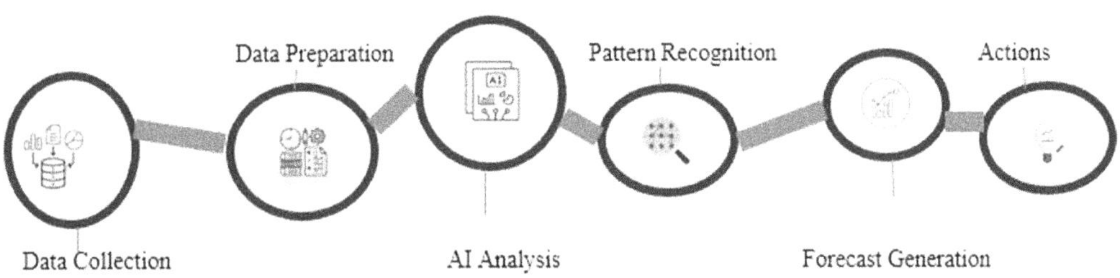

FIGURE 1.10 The AI demand forecasting process.

Demanding forecasting approaches are follows:

- **Time Series Analysis:** A statistical technique that uses past data to uncover patterns, trends, and seasonality in turn to forecast demand.
- **Neural Networks:** Analyzes sales information to predict demand and utilizes ML to capture and uncover complex patterns within the data.
- **Ensemble Methods:** This approach combines multiple models, thus reducing the drawbacks of the individual models, improving overall prediction accuracy.

1.1.2.9 NextGen AI bots:

NextGen AI bots help the farmers reach their customers and sellers instantly. They get quick help and guidance on the real-time problems. The vitals of the crop are monitored remotely, which include pest identification, control, and management of other sustainable factors. This 24/7 helpline offers support to the farmers, improving the overall outcomes (Gustavo, 2018).

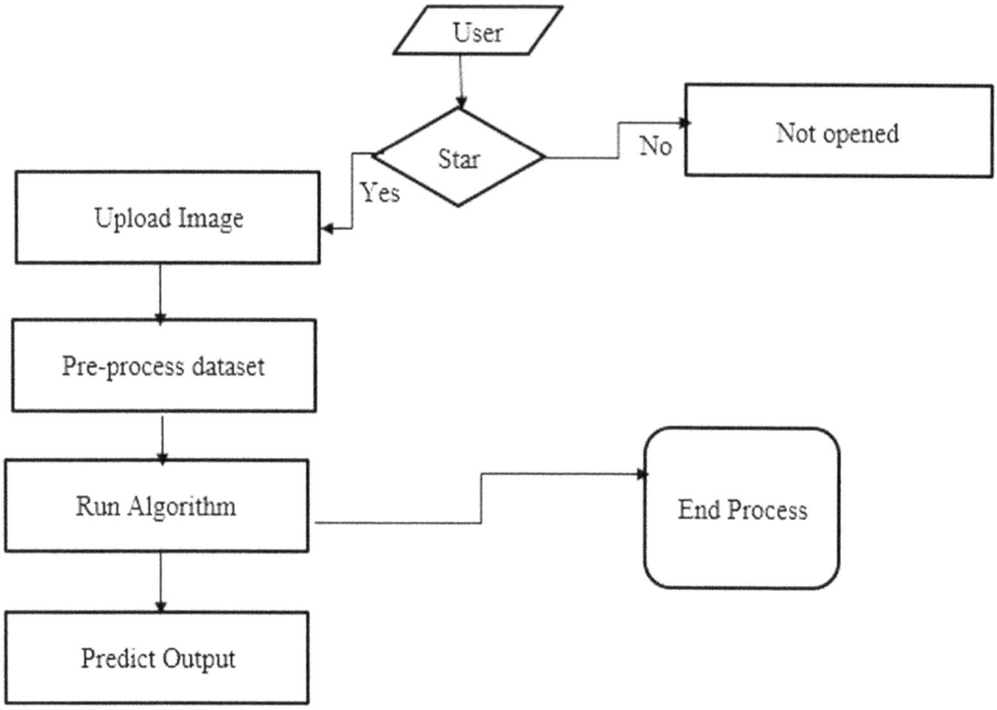

FIGURE 1.11 SA chat design.

Sustainable Agriculture (SA) Chat: The Chat Bot that can help farmers resolve issues involves uploading of crop images, with the application of a deep learning algorithm. The pests are identified and control mechanisms are suggested. The farmers can also ask queries in case problems related to crops.

AgronomoBot is a chatbot whose applications are seen in agricultural sensors. For use in agriculture, field conditions data like weather, air moisture, and soil conditions must be easily available. Agricultural management systems will assist experts and farmers by providing this information (shown in Figure 1.11), in turn, enabling them to plan strategies for decision-making.

AgronomoBot is programmed to look for and display information collected from a Wireless Sensor Network deployed in a vineyard environment (Frey, 2017). In addition, the IBM Watson cognitive services platform can be utilized to improve user experience through the integration of natural language processing features in interactions, thus facilitating intention recognition.

Additionally, improvements in AgronomoBot enabling features like messaging, speech and voice recognition,, image recognition, can provide additional insights.

This bot is expected to handle large volumes of data, avoid the risks, efficient pest management, and optimum resource and capacity utilization in SA (Bouteska, 2024).

1.1.2.10 NextGen AI open agents in SA Research Platforms

These platforms partner with agricultural companies who have incorporated AI to share the insights, they encourage innovation and ensure industrial development (Becker, 2000). The usage of AI tools in agriculture includes adaptation of new technology and the intention to innovate for a sustainable future.

TABLE 1.2
Comparison Between Traditional Method and AI-Enhanced Approach of Agriculture

Extension Service Area	Traditional Method	AI-Enhanced Approach	Estimated Improvement
Crop monitoring	Manual field inspections	Satellite imaging and AI analysis	90% time-saving, 95% accuracy
Pest management	Scheduled pesticide application	AI-powered predictive pest modelling	60% reduction in pesticide use
Yield prediction	Historical data and experience	Machine learning models using multiple data points	85% prediction accuracy
Resource allocation	Uniform application of resources	Precision agriculture with AI-optimized resource use	30% reduction in water and fertilizer use
Farmer training	In-person workshops and printed materials	AI-powered personalized learning platforms	70% increase in knowledge retention

1.2 APPLICATIONS OF AI: A COMPARISON

To understand the importance of AI in agriculture, it is essential to draw a comparison between the traditional and the AI approaches. An overview of the role of AI in transforming agriculture is presented in Table 1.2.

From the above table, it is clear how AI can transform the agricultural industry. The improved outcomes can substantially improve the life of a farmer and thus the country's economy

1.3 CHALLENGES IN AI ADAPTION

Data-Related Challenges

- o **Data Quality and Accessibility:** Ensuring high-quality, consistent, and accessible data often becomes difficult, especially while dealing with teams of various skill levels.
- o **Data Privacy and Security:** Protecting sensitive industrial data necessitates stringent cybersecurity measures and adherence to regulations like GDPR.
- o **Data Governance:** Creating strong frameworks for data ownership and usage rights is essential, yet complex.

Technical Challenges

- o **Scalability and Performance:** Scaling AI models and infrastructure to handle large volumes of data while maintaining efficiency is a significant challenge
- o **Interoperability:** Ensuring system integration with IoT devices requires a standardized interface and protocols, which is not an easily achievable task.
- o **Model Interpretability:** Improving transparency in AI decision-making is vital for building user trust, especially in regulated industries

Human and Organization Challenges

- o **Resistance to Change**: The work force resist AI adoption due to the fear of unknown and risk of losing jobs to AI.
- o **Skills Gap:** lack of skills, competencies, and experience in AI, data science, and domain knowledge hinders the implementation of AI

- o **Cultural and Leadership hurdles:** Organizations must foster a culture of innovation and data-driven decision-making, supported by servant leadership.
  Off late these risks are mitigated by separating the human work force from working with robots. They are placed in separate workspaces. These are practically difficult due to the collaborative work setting. They are called cohorts or collaborative robots (Pandya, 2020).
- o **Data overload:** Filtered and unfiltered data are used for both machine learning and data analytics. Large amount of data overwhelm the systems and slow down the decision making process.
- o **Cost:** The AI technology allows organizations to ensure their operations are both effective and cost-efficient.
- o **Achieve Balance between Human Super Power and NextGen AI Robots Roles:** Revisit and an improved thought process is required to achieve the balance between human super powered brains to the human-created NextGen AI SA chatbots. Human-centered AI solutions are need of an hour to shatter the misconceptions.

1.4 DIGITAL AGRICULTURE AI FOR NEXT GENERATION

1.4.1 AI-Powered Autonomous Drones

Imagine a scenario where fruits plucked from the trees are directly delivered to the consumers. AI-powered drones can do this for us. AI-powered autonomous drones are set to reshape last-mile delivery, making it faster, more efficient, and cost-effective. As technology advances, these drones could become a common sight, handling deliveries with precision and speed that were once unimaginable

The future of farming is changing quickly, thanks to AI. Smart farming is leading the way, with new tech emerging. Kubota North America Corp. bought Bloomfield Robotics Inc., (Wakchaure, 2023) showing AI's big role in farming.

Bloomfield's SaaS uses special cameras on Kubota tractors. It checks plant health, maturity, and fruit size. This helps farmers pick the best time to harvest and saves on labor costs.

The U.S. Department of Agriculture is investing in AI for specialty crops and organic farming. This move shows the government's support in different countries for AI in farming. We'll see more use of data and robots in farming soon.

Microsoft is backing Farmland LP, a big step towards AI in farming. Farmland LP uses satellites, electronic monitoring, and automated harvesting. This shows AI can make farming bigger and better (Ahir, 2020). It also helps fight climate change and supports green farming.to fully harness the potential of AI-based DSS in Industry 4.0. Organizations must

- Invest in robust data infrastructure and governance.
- Prioritize transparency and ethical considerations in AI design.
- Improve the resources skills, competencies, experience, and expertise in AI and promote shared clarity and collaboration.
- Inculcate a culture of adaptability and innovation, driven by visionary leadership.

The integration of AI-based DSS promises significant advancements in resource management and operational efficiency with successful deployment, fostering a competitive and sustainable industrial ecosystem.

1.4.2 Research and development efforts of Open AI

Open AI contributes significantly to Research and Development in areas requiring collaboration between the Human and Artificial intelligence. This team develops clear AI systems that collaborate with people (Glaroudis, 2020). Experts from the areas like computer science, cognitive studies and social sciences collaboratively work on Open AIs requiring interdisciplinary expertise (Kundalia, 2020).

Focus Points of Open AIs Research on Human AI Cooperation

Natural Language Processing (NLP): Language models such as GPT-3 (Generative Pre-trained Transformer 3) are developed by Open AI. These are designed to prompt text that closely resembles human conversation. These are used in content creation, chatbots, and language translation.

Computer Vision: In this area, OpenAI aims to develop algorithms capable of analyzing visual data and extracting meaningful insights. This includes projects like DALL-E, where images are generated using ANN with a feed of textual descriptions (Li, 2021).

Reinforcement Learning: Algorithms that can learn through trial and error and work parallelly with human workforce are created by Open AI. An example of this is OpenAI Five, a team of AI players designed for the video game Dota 2. The power of Human–AI collaboration is evident in several innovative technologies developed by OpenAI.

A Few Examples of OpenAI–Human Collaboration

1. **GPT-3:** This language model can generate text just like human notes with impressive accuracy. The largest language models are developing and suitable to different use cases.
2. **DALL-E:** it has the capacity to generate images from text prompts. The artificially created faces and scenarios may not exist in real life. For example, " A couch that looks like a Mango". This is an illustration of how the human creativity can be given a face.
3. **OpenAI Five:** DOTA 2 is a popular multiplayer online war arena, Open AI Five is group or team of AI agents capable of playing this by using the concept of reinforcement learning. where the past mistakes in the game help improve the future strategies in the game. It is an example of human–AI collaboration (Hong, 2001).

Making AI systems Clear, Understandable, and Explainable

The importance of having transparency and interpretability in AI systems is recognized by the Open AI. The organization is committed to creating AI systems that are transparent and explainable, enabling humans with knowledge of the system works and making informed decisions (Chen, 2001).

OpenAI has created a pathway to a numerous tools and methods to enhance transparency and capacity to interpret the outcome of the AI system which include the following:

Attention Mechanisms: This is a type of AAN architecture which supports humans to understand the way in which the model processes any information. The accuracy of such models are improved by detecting biases if any and removal of it on continuous basis.

Counterfactual Explanations: This is a technique which reasons the decision made by the model. This involves generating "what-if" scenarios to show how changing the input data would affect the model's output.

Human-in-the-Loop Approach: This technique involves humans, who provide feedback which is in turn fed into the AI model for training. This helps the more improve its accuracy

1.5 CONCLUSION

The NextGen AI solutions and futuristic technology innovations are revolutionizing the agriculture and improving the crop management across the globe. The concepts of sustainable farming, artificial

Intelligence, or the crop management systems are not hyped but are instrumental and essential in remodeling the agricultural landscape.

The data pipeline and continuous data models will help farmers to get reliable and accurate updates. An integrated approach using NextGen AI solutions seems promising as it can make farming not just productive but also profitable while staying sustainable.

In conclusion, the breakthrough in Open AI is redefining the research and development in Agriculture, the Ai and Human collaboration has set a standard for SA insisting on clear, understandable, and explainable models. This leads to better decision making, improved efficiency, and advanced creativity and innovation. This shift is enabling academicians, industry practitioners, and government bodies to collaborate more effectively, leading to improved outcomes.

REFERENCES

Adamchuk, V. I., Hummel, J. W., Morgan, M. T., and Upadhyaya, S. K. (2004). On-the-go soil sensors for precision agriculture. *Computers and Electronics in Agriculture*, 44(1), 71–91, ISSN 0168 1699. https://doi.org/10.1016/j.compag.2004.03.002. (www.sciencedirect.com/science/article/pii/S0168169904000444).

Ahir, K., Govani, K., Gajera, R., and Shah, M. (2020). Application on virtual reality for enhanced education learning, military training and sports. *Augmented Human Research,* 5, 7.

AlZubi, A. A., and Galyna, K. (2023). Artificial intelligence and internet of things for sustainable farming and smart agriculture. *IEEE Access*, 11, 78686–78692.

Arlitsch, K., and Newell, B. (2017). Thriving in the age of accelerations: A brief look at the societal effects of artificial Intelligence and the opportunities for libraries. *Journal of Library Administration*, 57(7), 789–798.

Becker, A., Bar-Yehuda, R., and Geiger, D. (2000). Randomised algorithms for the loop cut set problem. *Journal of Artificial Intelligence Research*, 12, 219–234.

Bouteska, A., Sharif, T., Bhuiyan, F., and Abedin, M. Z. (2024). Impacts of the changing climate on agricultural productivity and food security: Evidence from Ethiopia. *Journal of Cleaner Production*, 449, 141793. https://doi.org/10.1016/j.jclepro.2024.141793

Chen, C.-J., Huang, Y-Y., Li, Y. S., Chang, C-Y., and Huang, Y. M. (2020). An IoT based smart agricultural system for pests detection. *IEEE Access*. 1, 6–10.doi: 10.1109/ACCESS.2020.3024891.

Chen, X., and Van Beek, P. (2001). Conflict-directed back jumping revisited. *Journal of Artificial Intelligence Research*, 14, 53–81.

Feng, P., Wang, B., Liu, De L., Waters, C., and Yu, Q. (2019). Incorporating machine learning with biophysical model can improve the evaluation of climate extremes impacts on wheat yield in south-eastern Australia. *Agricultural and Forest Meteorology*, 275, 100–113, ISSN 0168 1923, https://doi. org/10.1016/j.agrformet.2019.05.018.

Frey, C. B., and Osborne, M. A. (2017). The future of employment: How susceptible are jobs to computerisation? *Technological Forecasting and Social Change,* 114, 254–280.

Gandhi, M., Kamdar, J., and Shah, M. (2020). Preprocessing of non-symmetrical images for Edge detection. *Augmented Human Research,* 5, 10, doi: 10.1007/s41133-019-0030-5.

Glaroudis, D., and Iossifides, A. (2020). Pericles chatzimisios, survey, comparison and research challenges of IoT application protocols for smart farming. *Computer Networks*, 168, 107037, ISSN 1389-1286, https://doi.org/10.1016/j.comnet.2019.107037.

Gustavo, M., Leonardo, C., Ícaro, S., and Carlos, C. (2018). AgronomoBot: A smart answering Chatbot applied to agricultural sensor networks. *14th International Conference on Precision Agriculture, Montreal,* Quebec, Canada. June 2018, 6–12.

Hong, J. (2001). Goal recognition through goal graph analysis. *Journal of Artificial Intelligence Research*, 15, 1–30.

Idoje, G., Dagiuklas, T., and Iqbal, M. (2021). Survey for smart farming technologies: Challenges and issues. *Computers & Electrical Engineering*, 92, 107104, doi:10.1016/j. compeleceng.2021.107104.

Indu, A. S. B., Bhardwaj, A., and Ibrahim, W. (2022). Optimization of pesticides spray on crops in agriculture using machine learning. *Computational Intelligence and Neuroscience*, 6-9 2022. https://doi.org/10.1155/2022/9408535.

Intergovernmental Panel on Climate Change (IPCC). Food Security and Food Production Systems. In: *Climate Change 2014 – Impacts, Adaptation and Vulnerability: Part A: Global and Sectoral Aspects: Working Group II Contribution to the IPCC Fifth Assessment Report*. Cambridge University Press; 2014, 485–534.

Jalal, A., de Oliveira Junior, J. C., Ribeiro, J. S., Fernandes, G. C., Mariano, G. G., Trindade, V. D. R., and dos Reis, A. R. (2021). Hormesis in plants: Physiological and biochemical responses. *Ecotoxicology and Environmental Safety*, 207, 111225, ISSN 0147-6513, https://doi.org/10.1016/jecoenv.2020.111225.

Kang, Y., Ma, X., and Khan, S. (2014). Predicting climate change impacts on maize crop productivity and water use efficiency in the loess plateau Irrigat. *Drain*, 63(3), 394–404, 10.1002/ird.1799.

Kumar, P., Singh, A., Rajput, V. D., Yadav, A. K. S., Kumar, P., Singh, A. K., and Minkina, T. (2022). Role of artificial intelligence, sensor technology, big data in agriculture: Next-generation farming. In Pradeep Sharma, Dinesh Yadav, Rajarshi Kumar Gaur (eds.). *Bioinformatics in Agriculture, Academic Press*, 625–639. ISBN 9780323897785. https://doi.org/10.1016/B978-0-323-89778-5.00035-0.

Kundalia, K., Patel, Y., and Shah M. (2020). Multi-label movie genre detection from a movie poster using knowledge transfer learning. *Augmented Human Research*, 5(2020), 11.

Li, J., Li, M. M., and Sun, L. P. (2017). Polarization maintaining microfiber-based evanescent wave sensors. *Acta Physica Sinica*, 66(7), 191–200. https://doi.org/10.7498/aps.66.074209.

Ma, Y.Q., and Sun, X. `(2020). Intelligent agricultural machinery equipment and technology. *Agricultural Equipment & Technology*, 46(1), 4–6.

Mahibha G, and Balasubramanian P. (2023). Impact of artificial intelligence in agriculture with special reference to agriculture information research. *Current Agriculture Research*, 11(1). doi : http://dx.doi.org/10.12944/CARJ.11.1.25

Maningas, R. V., Perez, V. O., Macaraig, A. J., Alesna, W. T., and Villagonzalo J. (2005). *Electronic information dissemination through the farmers' information and technology services (FITS)/Techno pinoy program. bringing information and technology within the reach of the farmers*. In Proceedings of the AFITA Conference. http://jsai.or.jp/afita/afita-conf/2000/ part08/p231.pdf, 2000.

Pandya, R., Nadiadwala, S., Shah, R., and Shah, M. (2020). Buildout of methodology for meticulous diagnosis of K-complex in EEG for aiding the detection of Alzheimer's by artificial intelligence. *Augmented Human Research*, 5, 1–8.

Qazi, S., Khawaja, B. A., and Farooq, Q. U. (2022). IoT-equipped and AI-enabled next generation smart agriculture: A critical review, current challenges and future trends. *IEEE Access*, 10, 21219–21235, doi: 10.1109/ACCESS.2022.3152544.

Sahoo, K., and Sethi, N. (2012). Investigating the impact of agriculture and industrial sector on economic growth of India. *OIDA International Journal on Sustainable Development*, 5, 11–21.

Wakchaure, M., Patle, B. K., and Mahindrakar, A. K. (2023). Application of AI techniques and robotics in agriculture: A review. *Artificial Intelligence in the Life Sciences*, 3, 100057, ISSN 2667-3185, https://doi.org/10.1016/j.ailsci.2023.100057.

Wihardi, Y., and Nugraha, E. (2020). Development mobile learning for vegetable farming in Indonesia based on mobile cloud computing. In *2020 6th International Conference on Science in Information Technology (ICSITech)*, Palu, Indonesia, pp. 6–10, doi: 10.1109/ ICSITech49800.2020.9392074.

Wakchaure, M., Patle, B. K., and Mahindrakar, A. K. (2023). Application of AI techniques and robotics in agriculture: A review. *Artificial Intelligence in the Life Sciences*, 3, 100057, ISSN 2667-3185, https://doi.org/10.1016/j.ailsci.2023.100057.

Gandhi, M., Kamdar, J., and Shah, M. (2020). Preprocessing of non-symmetrical images for edge detection. *Augmented Human Research*, 5(1), 10.

2 AgroBot

A Domain-Specific Agricultural Knowledge System Using Retrieval-Augmented Generation

Rachana M S, Shridevi A Desai, Roopashree S, and Shashidhar R

2.1 INTRODUCTION

Farming plays a crucial role in the world economy, greatly supporting food availability, job creation, and international commerce. In the last 20 years, the farming industry has experienced substantial development driven by cutting-edge technologies and growing production (Vizniuk et al., 2024). The data-driven approach promises to improve crop quality, resource allocation, and overall agricultural management. The key to the success of precision agriculture is the provision and dissemination of agricultural information to farmers, enabling them to make well-informed choices regarding agricultural practices. Several obstacles exist that occur in farming globally; these include climatic conditions, market volatility, and many others; there is a high need for improved facilities in helping in decision-making within the agricultural sector. Through this Chabot, the revolution is being availed not only in making all the information that the farmers may need regarding farming, but even with the options that may be tailored to suit the needs of every farmer (Balpande et al., 2024).

Agriculture ranks among the most crucial and, at the same time, promising sectors where AI technologies are making a significant impact. Agriculturists ensure proper resource utilization while boosting crop production and awareness regarding environmental-friendly practices. They combat climate change challenges, create protective boundaries for ecosystems, and help enhance agriculture progress to meet the needs of the future. In this aspect, AI-driving agriculture is rapidly gaining momentum as a vital tool for improving farming practices and promoting sustainability. A wide variety of AI-based methods have been employed for plant disease diagnosis, pest control management, crop monitoring, boosting crop yields and many others, fostering education and decision-making processes. The agricultural sector is facing significant challenges as the global population grows, increasing the demand for food. Many farmers lack access to precise, location-specific knowledge and continue to rely on traditional farming methods that may no longer be effective due to changing environmental and market conditions. The agriculture connection framework is designed to address this issue by providing farmers with data-driven insights. The platform leverages advanced technologies to analyze large datasets and offer practical recommendations. Utilizing LLMs for agriculture encounters numerous critical issues concerning the domain and knowledge specifics. Thus, for example, agricultural science utilizes specialized vocabulary and concepts (e.g., soil nutrient cycles, integrated pest management) that generic LLMs may not fully capture unless they are fine-tuned on agriculture-specific datasets. Accurate understanding of context-specific information (like regional farming practices or environmental factors) requires domain adaptation, which may be challenging given the high diversity within agricultural practices around the world. Data quality and availability raise another challenge because agricultural information is gathered from multiple origins (textbooks, research articles, field reports, and sensor data) and vary in quality, formats, and

 DOI: 10.1201/9781003613510-2

granularity. Large, high-quality, and well-annotated datasets specific to niche agricultural areas are often limited in availability, especially when compared to the more abundant and comprehensive datasets found in broader domains (e.g., general web data or news). Finally, when an LLM is applied to specialized domains like agriculture, errors or misinterpretations in the generated content may lead to practical problems, such as incorrect advice on crop management or resource allocation. As a result, farmers and agricultural experts may be sceptical of automated systems, especially if the outputs are overly generic (Samuel et al., 2025).

It aims to develop a comprehensive agriculture-specific retrieval augmented generation (RAG)-LLM framework called AgroBot which is a centralized knowledge base for agriculture. As an initial step toward this objective, this chapter provides a structured approach to integrating RAG over LLMs into agricultural applications, demonstrating the complete workflow, beginning with data collection and pre-processing to embedding generation, retrieval using vector databases, and retrieval-augmented generation. It shows how existing technologies can be adapted to a specialized domain, serving as a proof of concept that can spur further research and development. The contribution to the work includes the following:

1 The RAG LLM framework bridges the knowledge gap by connecting external knowledge databases with practical applicability for farmers.
2. Using an agriculture-specific database using an extensive corpus of dataset to facilitate efficient information retrieval in the agricultural domain.
3. Designed the RAG framework for agricultural knowledge generation, which uses a combination of cutting-edge AI technologies to deliver contextually relevant and accurate agricultural information. The text extraction is performed using generative AI, and the resulting embedding is stored in Facebook AI Similarity Search (FAISS) format. This ensures efficient similarity searches and retrieval along with precise access to agricultural knowledge.
4. To select a benchmark generative AI model for the RAG, Llama 3.2 model is used and assessed their effectiveness in generating responses related to farming and agriculture. The evaluation parameters include accuracy, relevance, response time, and ability of models to respond to complex agricultural queries.

This chapter presents a structured overview of the AgroBot system, a domain-specific agricultural knowledge solution using RAG. It begins with an abstract and introduction, followed by a literature survey highlighting research gaps. The methodology section outlines the AgroBot pipeline, while the RAG-LLM framework details embedding, retrieval, and response generation. Results compare LLM-only and RAG-based responses, and evaluation metrics demonstrate improved performance using BERT Score. The chapter concludes with key findings and references for further research.

2.2 LITERATURE SURVEY

Jeya et al. (2024) use Google's Gemini LLM, fine-tuned with user-generated data from CSV files containing dealer information. Location accuracy is enhanced using Google Maps API and Serp API to fetch and store real-time dealer data in a vector database. It analyses climate, soil type, and moisture levels to suggest the most suitable crops. It utilizes the fine-tuned Gemini model for precise, location-specific recommendations. It describes the Crop Recommender Model which analyses climate, soil type, and moisture levels to suggest the most suitable crops. Agriconnect represents a breakthrough in smart agriculture, using LLMs, real-time data, and AI-driven recommendations to empower farmers with actionable insights. However, it did not account. The system relies on user-generated CSV files for dealer details. If these files are incomplete, outdated, or inaccurate, it could affect recommendations. Climate and soil data are provided by users. There is no mention of integrating real-time satellite data or IoT-based sensors to enhance accuracy. The use of a vector database is efficient, but it's unclear how the system handles large-scale data from multiple users simultaneously. Server load management and response time for real-time crop recommendations are not discussed.

Silva et al. (2023) focus on prior research related to LLMs, their applications in various domains, and their relevance to agriculture and education. The paper references several studies showcasing LLMs' ability to excel in various fields such as medicine, law, psychology, and education. Nori et al. (2023) assessed Generative Pre-trained-4 using medical proficiency examinations, showing that despite not being trained specifically for medicine, the model performed well. Nunes et al. (2023) examined GPT-3.5 and GPT-4 on Brazil's ENEM exam, finding that chain-of-thought (CoT) prompted improved performance. Singhal et al. (2023) introduced Med-PaLM 2, a medical LLM that combines domain-specific fine-tuning with ensemble refinement (ER), a method similarly applied in the paper. The study incorporates RAG to improve response accuracy, a technique proposed by Lewis et al. (2020) that allows LLMs to collect relevant data prior to generating responses. ER, proposed by Singhal et al. (2023), improves accuracy by refining answers based on multiple generations. The paper builds on self-consistency and chain-of-thought (CoT) prompting strategies (Wei et al. 2022), which enhance LLM reasoning. Limited research on LLMs specifically for agriculture – Most prior studies focus on machine learning in agriculture, but few explore the role of LLMs in agronomic decision-making. Lack of real-world testing of LLMs for farmers – Studies primarily evaluate models on exams, but do not assess the performance of these models in advisory roles for farmers. RAG and ER techniques have been used in medical AI but not extensively in agriculture. The study applies RAG and ER to agricultural contexts, an area not widely explored. Limited research on multilingual AI in agriculture – Previous studies does not investigate how well LLMs handle region-specific dialects and farming practices. The literature review provides a solid foundation for the study by linking prior advancements in AI, education, and agriculture to the potential of GPT-4 in agronomy. However, it highlights that LLMs in agricultural advisory roles remain underexplored, paving the way for the current research.

Chia et al. (2024) provide an overview of LLMs, their architectures, and their applications in agriculture. The paper explores two primary strategies for improving LLM performance. Fine-tuning involves adapting a pre-trained LLM using domain-specific data to generate highly accurate and task-specific responses. It requires extensive computing resources and retraining for updates. RAG enhances LLM responses by retrieving relevant external information before generating text and allows the model to dynamically access external knowledge, reducing the risk of hallucinations may generate more complex outputs and making response interpretation challenging. Balaguer et al. (2024) compared RAG vs. fine-tuning in agriculture. The paper reviews various evaluation methods used to measure LLM effectiveness: Hugging Face Ranking System uses public voting, MT-Bench, and MMLU scores. GLUE & SuperGLUE – uses standards for natural language understanding. General AI Assistant Benchmark (GAIA) evaluates how well models handle queries. The paper does not explore real-world adoption of LLMs for irrigation optimization. It lacks a discussion on multilingual capabilities of LLMs in agricultural settings. There is limited focus on integrating LLMs with IoT-based smart irrigation systems.

Balaguer et al. (2024) compared RAG and fine-tuning techniques for incorporating domain-specific knowledge into LLMs. The authors propose a pipeline that integrates both approaches and analyse their tradeoffs across multiple models, including GPT-4, Llama2-13B, and Vicuna. The study is particularly focused on the agriculture sector, where the uptake of AI has been sluggish because of the absence of industry-specific training data. The paper examines whether LLMs can provide geographically specific insights to farmers by leveraging fine-tuning and RAG. The paper "A Comprehensive pipeline for RAG and Fine-Tuning" presents a well-structured pipeline that includes data acquisition, Q&A generation, information retrieval, fine-tuning, and evaluation. It evaluates multiple LLMs across different techniques. The study focuses on agriculture, an underexplored domain for LLMs. The research highlights the importance of context-specific AI applications. The paper evaluates GPT-4, Llama2-13B, and Vicuna on their ability to incorporate agricultural knowledge. It provides numerical performance indicators for each model. The paper explicitly discusses the pros and cons of RAG and fine-tuning. Fine-tuning: More accurate but expensive and requires frequent retraining. RAG: more flexible, cheaper, and enables real-time updates but can generate hallucinations.

Muhammad Arslan et al. (2024) investigated RAG with LLMs, addressing the shortcomings of LLMs in managing domain-specific queries, through the inclusion of fetching external data for the generative workflow. The study identifies research gaps and future directions while offering a thorough review of RAG implementations spanning diverse tasks and fields. Research papers on RAG applications were gathered via a structured literature review using Google Scholar. Chosen studies were categorized into discipline-based (e.g., medical, finance, education) and task-based (e.g., QA, text generation, information retrieval) categories that examined how RAG research changed between 2020 and the beginning of 2024. RAG grounds responses in retrieved evidence, which lessens hallucinations in LLM outputs. RAG is utilized in the fields of medicine, finance, education, and law. Medical, financial, legal, and educational fields all benefit from RAG's broad applicability, which improves applications including QA, text summarising, and decision support. Improved real-time information processing dynamically integrates outside knowledge, thus LLMs are more helpful for modern and context-specific projects. Many RAG papers are unpublished and lack peer review, so their validity is called into doubt depending on preprint studies. RAG implemented with closed-source LLMs can be costly computationally, restricting accessibility. Processing big, varied sets could cause delays and can result in scalability problems. The survey does not explore architectural-specific improvements as technical implementation details are insufficient. Accuracy is an issue as RAG answers still can be deceptive and call for more proof. Handling sensitive biomedical and legal data raises ethical questions about privacy. In conclusion and with reference to future work, the paper draws attention to research gaps particularly in fields like entity linking and real-time RAG implementations and calls for more research on ethical consequences of RAG in sensitive areas, dataset scalability, and economical application.

Biplov et al. (2024) presented RAG Chabot's coupled with a hybrid deep learning model to classify medicinal plants and offer ayurvedic agricultural insights. Targeting farmers with medicinal plant identification and usage recommendations, the system supports both Nepali and English. Combining local and global feature extraction, the technique uses DeiT (Data-efficient Image Transformer) and VGG16 CNN models, trained on a 4161 training and 893 test image set including Indian Medicinal Plants. It uses the OpenAI API to implement RAG in producing plant insights. English and Nepali bilingual support uses Google Translate API. High accuracy is achieved as the accuracy of the DeiT + VGG16 hybrid model is 96.75%, which is higher than that of DeiT (95.97%) and VGG16 (90.26%). The RAG-based Chabot enables the user to gain economic and medical information enhancing plant classification and retrieval. Bilingual and offline features support dissemination of information while reducing the technological constraints in the countryside. A few visually alike plants are misled by the model leading to misclassification problems. Misinformation risks are expected as RAG sometimes produces inaccurate conclusions regarding plant use. Language translation issues arise as relying on Google Translate API, causing Nepali translations to be error-prone. The present model does not cater to all medicinally valuable plant species limiting the scalability. The project focuses on developing a real-time mobile application with an enhanced chatbot user interface and integrated IoT support for efficient plant tracking. Future development involves increasing plant species coverage and improving model precision with the help of sophisticated deep learning methods.

Marvin Kasting et al. (2024) assessed LLMs for question-answering and information retrieval in agriculture based on a newly created BVL QA Corpus 2024. This paper also presents a RAG system for improving information retrieval in the agricultural industry. It created the BVL QA Corpus 2024, which includes 41,791 passages of 4,952 agricultural texts. It employed LLMs

(GPT-4, GPT-3.5, LeoLM, and Mistral-7B) to produce and score question-answer pairs. It compared models based on BLEU, ROUGE, METEOR, and SAS metrics. It installed a RAG system for more contextual retrieval and accurate answers. RAG systems excel compared to isolated LLMs for question-answering because they enhance precision and relevance. New benchmark dataset offers a normalized test framework for agricultural LLM use cases. It confirmed LLM capabilities in extraction and synthesis of agricultural knowledge. Open-source LLMs lack agricultural domain experience, underperforming GPT-3.5/GPT-4. Naïve RAG implementation has its limits, with limitations on metadata extraction and document chunking. Benchmark dataset remains untested in the real world using synthetic testing. Upcoming studies will center on enhancing RAG retrieval approaches, refining embedding's, and checking for factual correctness.

Banerjee et al. (2024) discussed how LLMs and AI are revolutionizing agriculture through increased productivity, precision agriculture, and data-informed decision-making. It identifies the potential of RAG to enhance agricultural information retrieval and assist farmers with real-time, context-aware information. AI-powered analytics maximize crop rotation, irrigation, and soil health monitoring. Decision support automation is achieved as LLMs aid in pest identification, disease management, and yield estimation. AI assists in price forecasting, supply chain logistics enhancement, and minimizing post-harvest losses. RAG-boosted LLMs can offer localized agricultural guidance to farmers enhancing language accessibility. LLMs need high-quality, domain-specific data, which might be scarce. Computational expense is high as AI-powered agriculture systems consume high computational capacity, making adoption expensive. LLMs can produce misleading or incorrect farming advice because of model biases. Small farmers do not have digital infrastructure to harness AI solutions.

Selva Kumar et al (2024) proposes an AI-based precision agriculture solution that combines the use of YOLOv8 for disease detection and RAG for context-aware diagnosis in coffee cultivation. The solution is designed to correct hallucinations in LLMs and improve real-time coffee leaf disease identification and remediation processes. It leverages the use of YOLOv8 for real-time object detection to determine coffee leaf diseases (e.g., *Phoma, Miner*) and utilizes the use of GPT-3.5 for disease identification and treatment suggestions. This work integrates RAG to retrieve timely, domain-specific agricultural information to minimize LLM inaccuracies. It creates a hybrid workflow, which combines image analysis, NLP, and external knowledge retrieval. Precise disease identification is achieved as YOLOv8 enhances speed and accuracy of detection. Context-sensitive diagnosis is facilitated as RAG mitigates LLM hallucinations through retrieval of real-time agrarian knowledge. The system encourages environmentally friendly agriculture through the reduction of pesticide dependency. It developed to accommodate real-time surveillance and dataset enlargement across crops which makes it scalable and adaptable. Data dependence is huge as this needs huge, labelled datasets to train YOLOv8 and GPT-3.5. Execution of RAG and deep learning models requires high computing capability. While RAG decreases hallucinations, LLM-predicted findings still need the verification of specialists. The system is currently limited to a few coffee diseases and requires expansion to Arabica and Robusta varieties. Future work should focus on scalability, dataset enhancement, and real-world validation with farmers.

2.3 METHODOLOGY

2.3.1 Dataset Overview

This study utilizes RAG-enhanced language models, incorporating context-specific questions and answers sourced from datasets related to India, a major global leader in agriculture. The agricultural sector serves as the representative industrial domain for this research. The dataset compiled for the AgroBot system encompasses diverse formats and content types, such as regulatory guidelines, scientific publications, agronomic evaluations, and structured knowledge repositories. This section offers a comprehensive overview of the data utilized in developing the system.

India: Within India, the KVK (Krishi Vigyan Kendra) functions as a specialized extension for agricultural establishment, specifically dedicated to facilitating the localized adoption of agricultural research findings. The fundamental objective centers on establishing connectivity between agricultural practitioners and scientific investigation outcomes. This knowledge transfer centers deliver diverse extension support mechanisms for agricultural communities, including telephonic consultation infrastructure for addressing producer inquiries. The documented communication exchanges from these interactions constitute publicly accessible datasets available through the governmental data portal at https://data.gov.in/datasets_webservices/datasets/6622307. The following fields are present in each of the numerous sets of records: crop type, weather-related query type, government schemes, plant protection, timestamp, and growing season (rabi or kharif). The coverage extends across diverse domains, including agro-enterprises, optimal agricultural practices, aquaculture, sustainable forestry, organic cultivation, relevant governmental policies and schemes, post-harvest processing technologies, poultry farming, meteorological data, crop cultivation, essential agricultural inputs, and agricultural export guidelines, among others. Farmers, students, researchers, and anybody else with an interest in agriculture can all benefit from its wealth of information (Balaguer et al., 2024).

This chapter describes a methodology based on a structured pipeline, called AgroBot that automatically generates and assesses question-answer (QA) pairs for the creation of domain-specific co-pilots in agriculture. Data are first obtained from the Vikaspedia portal (https://vikaspedia.in/agriculture/), a comprehensive resource providing general information across a wide range of agricultural domains, as shown in Figure 2.1 of the proposed pipeline. This resource comprehensively covers domains such as crop cultivation, agricultural inputs and export strategies, insurance, agribusiness models, optimal farming practices, fisheries management, forestry, organic agriculture, pertinent governmental policies and programs, post-harvest technologies, poultry farming, and meteorological information, among others. Vikaspedia serves as an indispensable knowledge repository for farmers, students, researchers, and any individuals engaged in or interested in the agricultural sector. The specific data sources and document types employed in this study are clearly exemplified. There are clear examples of the data sources and document types used in this study. The first step in the process is gathering agricultural data, which serves as the pipeline's foundation. Together with carefully chosen texts from pertinent literature, the dataset includes test questions taken directly from the texts, paraphrased questions, actual user queries, noisy text samples, and external context chunks. All of these are intended to assess semantic comprehension and the system's capacity to manage intricate information requirements.

The development of RAG follows a structured methodology such as data collection, information extraction, question generation, and answer generation.

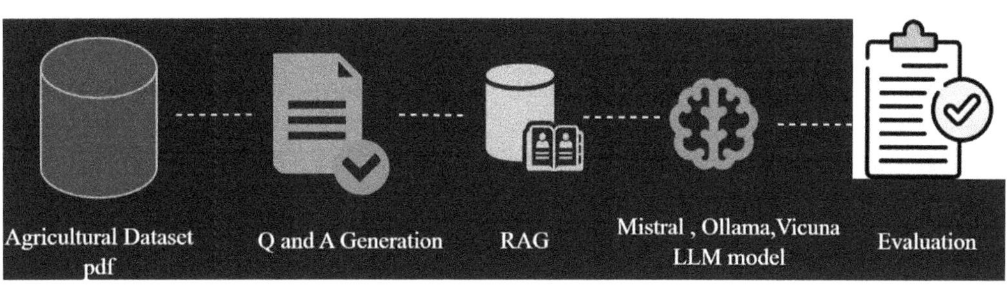

FIGURE 2.1 The proposed pipeline of AgroBot architecture.

2.3.2 Data collection

The pipeline's primary goal is to compile a diverse and carefully selected dataset that includes data relevant to agricultural crops. This capability facilitates the construction of both queries and corresponding responses, thereby establishing the groundwork for enhancing model precision and the relevance of generated outputs. The data sources chosen for this step offer reliable, high-quality information on the subject at hand. In agriculture, for instance, this includes databases of agronomist exams, scientific knowledge repositories, and agricultural and environmental government agencies. Furthermore, the extracted information must precisely align with the factual grounding furnished to the model.

2.3.3 Information Extraction

The efficacy of subsequent processing steps within this chapter is contingent upon the quality of information and textual structure derived from the assembled PDF documents. This presents a considerable challenge; however, as the primary design objective of the PDF format is consistent document rendering across diverse platforms, not seamless data extraction. The sections, subsections, and related content that make up a document's logical structure do not correspond to the underlying structure of a PDF file. Furthermore, we find that the formatting and layouts of documents that come from different sources are complicated and inconsistent, frequently displaying a variety of sidebars, tables, images, and page footers.

2.3.4 Question Generation

The fundamental objective of this computational module within the processing framework focuses on managing the inherent variability and non-deterministic characteristics exhibited by human linguistic expression during the question generation phase from source documents. It seeks to produce excellent, contextually based questions that faithfully capture the essence of the extracted text.

2.3.5 Answer Generation

The proposed methodology implements RAG, an innovative computational paradigm that synergistically integrates document extraction capabilities with neural synthesis mechanisms to produce superior textual outputs. This approach demonstrates exceptional efficacy when processing voluminous and multifaceted knowledge repositories, as it systematically identifies and extracts documentary evidence pertinent to specific inquiries, subsequently leveraging these materials to enhance the synthesis process. The computational workflow initiates through the identification of optimally relevant textual segments from the knowledge base corresponding to input queries, employing dense vector-based matching algorithms for document selection. These extracted materials function as foundational knowledge substrates for the downstream synthesis phase. Following successful passage identification, an LLM processes the interrogative input in conjunction with the extracted documentary evidence, generating contextually coherent responses. The synthesis procedure remains anchored to the informational boundaries established by the retrieved materials, ensuring the produced interrogative-response dyads exhibit accuracy, semantic validity, and substantive value. The implementation encompasses three sequential operational phases.

2.3.5.1 Embedding Generation and Index Construction

Computation of vector representations from segmented textual units extracted from portable document format repositories, utilizing transformer-based sentential encoding architectures. Subsequently, the Facebook AI Similarity Search (FAISS) infrastructure facilitates the construction of an efficient searchable vector repository, enabling rapid similarity computations and document retrieval operations.

2.3.5.2 Retrieval

Given a question input, its embedding is calculated and employed to retrieve the most appropriate chunks of text from the FAISS database. The process of retrieval is conducted through the similarity_search_with_score method, which is widely used in vector databases and libraries like FAISS. The method accepts two primary arguments: the query (input question) and top-k, which specifies the number of top-most similar documents to return. It performs a similarity search and returns the top-ranked text snippets along with their corresponding similarity scores, effectively identifying the most contextually relevant content for the query.

2.3.5.3 Answer Generation

An LLM architecture processes the interrogative input alongside the extracted documentary segments to synthesize comprehensive responses. The implementation leverages GPT-4, which receives contextual information retrieved from the vector similarity search infrastructure as supplementary input through a specifically engineered prompt template, facilitating the generation

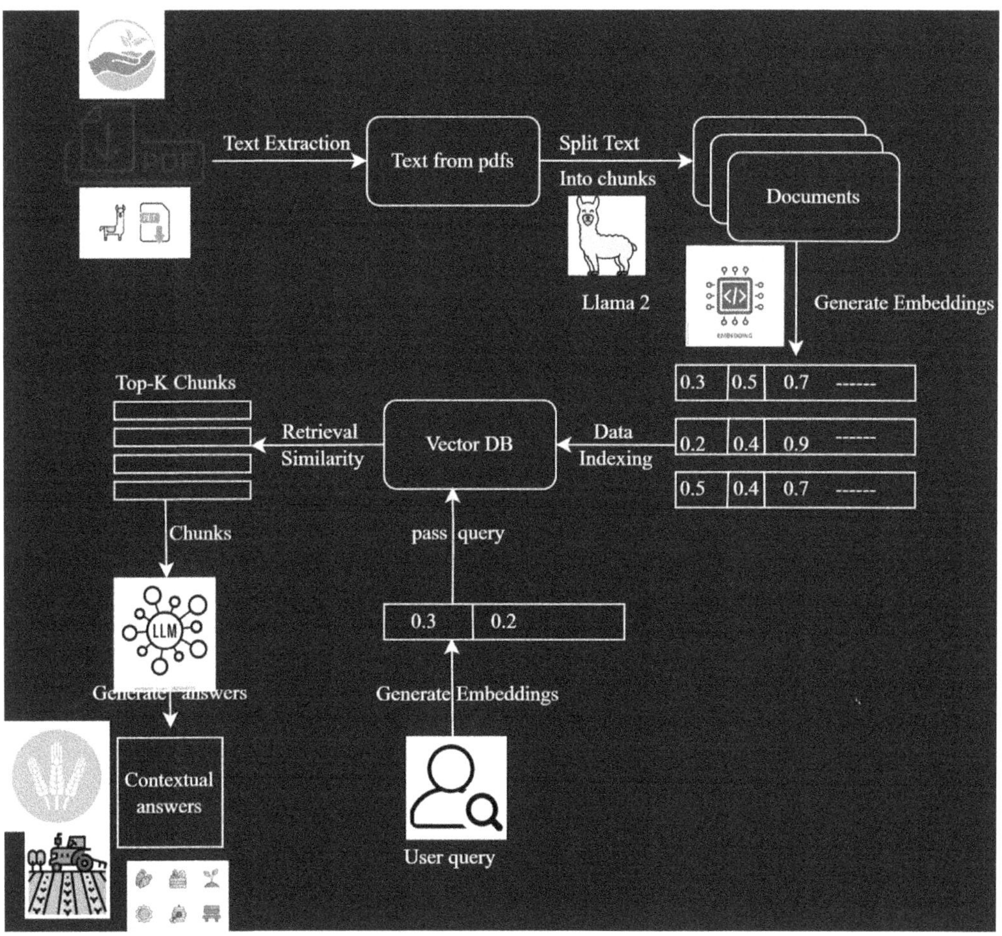

FIGURE 2.2 Pipeline of RAG-based system for contextual answer generation from agricultural PDFs.

of responses tailored to the target knowledge domain. This configuration enables the synthesis of outputs that demonstrate deep contextual understanding and domain-specific expertise by grounding the generative process in the retrieved documentary evidence stored within the FAISS-indexed repository. AgroBot integrates generative AI–driven embedding generation with a vector database, facilitating efficient and scalable semantic search. General workflow for information retrieval and generation encompasses query processing, embedding generation, database management, and retrieval-augmented response generation, as illustrated in Figure 2.2. Text is extracted from these PDFs using a text extraction tool, and then extracted text is divided into smaller chunks for better processing language model like Llama 3.2, Llama 2. Then, each text chunk is transformed into an embedding a numerical vector representation of the text. These embedding's are indexed and stored in a Vector DB. It generates embedding's for the query and passes these embedding to the Vector DB. The Vector DB compares the user query embedding's with document embedding's to retrieve the most relevant chunks (Top-K retrieval). These retrieved chunks are passed to a language model to generate contextual answers based on the information.

2.4 RAG-LLM FRAMEWORK

2.4.1 Embedding Models: Conversion of Chunks into Embedding

To manage the extensive information contained in these agriculture books, the documents are divided into smaller, manageable chunks. This chunking process is essential to create a meaningful embedding and ensure efficient storage and retrievals. Each chunk was carefully crafted to encapsulate a specific topic or sub-topic, ensuring that the granularity was appropriate for educational use. The varied chunk sizes depend on the complexity and cohesiveness of the content, typically ranges from a few sentences to a couple of paragraphs. This metadata helps to accurately situate each chunk within the broader structure of the book, making it easier to provide relevant and contextually accurate responses when users query the system. By ensuring that each chunk is appropriately sized and contextually rich, which aims to facilitate a more effective and engaging learning experience for users.

The embedding are compact vector representations of text that convey the semantic meaning of the chunks. Utilizing embedding models from Llama 3.2 to convert text chunks into embedding's. These models, trained on vast corpora of text, provide robust representations that facilitate effective similarity searches. By processing the chunks with these models, we generated embedding vectors that encapsulate the semantic context of the text. It is stored in the embedding vectors together with their associated metadata, which includes information such as chapter and section headings. This metadata is crucial for preserving context and ensuring that the retrieved information is relevant and accurately situated within the broader structure of the document. By maintaining a detailed metadata record, we enhance the capability of the Chabot to provide precise and contextually appropriate responses. This process of converting text chunks into embedding's is the crucial step in the project. It enables efficient storage and retrieval of information, allowing users to quickly access relevant and detailed agricultural knowledge. By using advanced AI models, we ensure that the embedding's are both accurate and rich in semantic content, supporting a more effective and engaging learning experience for users.

2.4.2 Database Lookup Using FAISS

The computed vector representations undergo preservation within a specialized indexing infrastructure utilizing the Facebook AI Similarity Search (FAISS) framework. This computational architecture facilitates efficient nearest-neighbor retrieval operations and vector space partitioning algorithms, rendering it particularly suitable for managing high-dimensional representations across large-scale corpora. This step ensures that embedding's are organized to allow rapid retrieval based on similarity measures. The choice of FAISS is driven by its scalability and performance in handling large volumes of vectors. This is crucial in the project, as it involves managing a significant amount of data derived from comprehensive agricultural texts. Further storing embedding's in FAISS allows us to efficiently manage and query the data. Upon user query, the system can rapidly search through the embedding's to find the most similar chunks, providing accurate and contextually relevant responses. This setup enhances the functionality and responsiveness of our agricultural-specific Chabot, offering improved user experience. Overall, using FAISS to store the embedding ensures to handle the large-scale data efficiently while maintaining high performance in similarity searches.

2.4.3 Prompt Construction and Sending Relevant Chunks to the LLM

The retrieved text chunks are combined into a structured prompt, which is then sent to an LLM using the RAG technique. The RAG approach combines text generation with searching and selecting the most relevant data out of a pre-indexed vector database. The user query and the relevant information are both given to the LLM. The LLM uses the retrieved information to create better responses. RAG allows the model to create responses based on the context obtained from the database. This reduces the risk of unpredictable or incorrect responses typical of generative models.

2.4.4 Response Generation and Knowledge Sharing through RAG-LLM Framework

The RAG-LLM framework is designed as an agricultural tool tailored for agriculture-related topics and implemented as a Chabot. This bridges the knowledge gap, connecting farmers with effective agricultural practices. As a virtual assistant, it uses the RAG-LLM framework relevant text chunks to provide explanations, answer questions, and offer insights into various agricultural and farming concepts. The user interface sends the query to the RAG system, which acts as the orchestrator, managing how the query is processed. The RAG converts the query into an embedding and submits it to the FAISS vector database containing pre-computed embedding's of text chunks from curated agricultural texts. The vector database performs a similarity search to recognize text portions that are contextually closest to the user query. To make sure that users receive responses even when relevant documents are unavailable, the system is equipped with a fall back mechanism. The RAG system evaluates whether relevant documents were retrieved. If yes, the retrieved chunks are combined with the original query to create a context-rich prompt for the LLM. Otherwise, the system falls back to the LLM's internal knowledge to generate a response. Response generation is performed through the LLM. The LLM analyses the assembled prompt—which combines the retrieved information segments with the user's query and generates a comprehensive and cohesive response. In case if no relevant documents were retrieved, the LLM uses its pre-trained internal knowledge to address the query. The generated response is sent back to the RAG system and then delivered to the user through the user interface. The accuracy and relevance of the final output largely rely on the LLM's capability to seamlessly combine and clearly present the information extracted from the retrieved segments (Parthiban and Adil, 2023; Dong, 2024).

2.5 RESULTS

Table 2.1 presents the results for six user queries extracted from multiple PDF documents. For each query, three types of responses were generated: a baseline response produced by the language model without external context (LLM-response), a RAG-based response utilizing relevant information retrieved from the documents (RAG-response), and the raw contextual data retrieved from a vector database, which served as input for the RAG model (Contextual response).

TABLE 2.1
Three Types of Responses: LLM Response, RAG Response, and Contextual Responses (Vector Database) Retrieved from a Database

Query	LLM response	RAG response	Contextual response
How do beetles kill the *parthenium*?	Beetles that feed on the *parthenium* plant (Parthenium hysterophorus) can be a significant factor in controlling its spread and growth. Here are some methods by which beetles can kill the *parthenium*: 1. Feeding damage: Beetles can feed on the leaves, stems, and flowers of the *parthenium* plant, causing damage and weakening it. As the plant becomes weaker, it is more susceptible to other stresses, such as disease and environmental factors. 2. Egg laying: Some beetle species lay their eggs directly on the *parthenium* plant that can cause the hatching of larvae that feed on the plant tissues. These larvae can cause significant damage to the plant by soft chewing on the leaves	Based on the given context, beetles kill *parthenium* through both adults and larvae feeding on the plant's leaves. The larvae feed on soft growing leaves, starting with the terminal and axillary buds and later the leaf blades, which check the plant's growth and flower production. The adult beetles voraciously defoliate the plant, cutting immature flowers and chewing on the soft tissues beneath them. Completely defoliated plants start to show die-back symptoms and eventually get killed	Both adults and larvae are capable of feeding on *parthenium* leaves. Larvae after hatching start to feed on soft growing leaves, first attacking the terminal and axillary buds and later the leaf blades, thus checking completely the plant growth and flower production. 1 10 * - Completely defoliated Plant; I *Parthenium* plants nipped in the bud Adults voraciously defoliate the plant. Immature flowers are cut by die beetles in an effort to chew soft tissues beneath the flowers
What are the benefits of brown manuring?	Brown manuring, also known as composting, is the process of breaking down organic matter into a nutrient-rich soil amendment. Here are some of the benefits of brown manuring: 1. Improves soil structure: Brown manure can help improve the texture and structure of soil by adding bulk density, increasing porosity, and improving water infiltration. This can lead to better drainage and aeration, which can benefit plant growth. 2. Increases soil fertility: The nutrients present in brown manure, such as nitrogen, phosphorus, and potassium, can help improve the fertility of soil. These nutrients can be made available to plants through decomposition	Based on the given context, the benefits of brown manuring can be summarized as follows: 1. Replacement of nitrogenous fertilizer: Brown manuring supplies nitrogen to component crop plants through biological nitrogen fixation, thus replacing a significant part of nitrogenous fertilizer. 2. Improved soil organic carbon content: The incorporation of legume-derived organic matter into the soil increases its organic carbon content, which improves the physicochemical and biological properties of the soil. 3. Soil conservation: Brown manuring can conserve soil moisture and reduce runoff and wind erosion, due to the increased soil organic matter	Legume brown manuring crops supply nitrogen to component crop plants through biological nitrogen fixation and thus, a significant part of nitrogenous fertilizer can be replaced. It increases organic carbon content of soil and improves the physicochemical and biological properties of the soil. Brown manuring can conserve soil moisture and reduce runoff and wind erosion. Enhanced soil fertility, lesser weed competition as well as other benefits accrued under brown manuring, leads to higher crop productivity and economic gain to farmers

TABLE 2.1 (Continued)

Query	LLM response	RAG response	Contextual response
Classify types of seeds	There are several ways to classify seeds based on their characteristics and properties. Here are some common methods: 1. Based on seed size: Small seeds (e.g., *mustard*), medium seeds (e.g., *sunflower)*, large seeds (e.g., *pumpkin*). 2. Based on seed shape: Round seeds (e.g., *peas*), oval seeds (e.g., *beans*), irregularly shaped seeds (e.g., *sesame*). 3. Based on seed coat: Smooth seeds (e.g., *corn*), rough seeds (e.g., *quinoa*), hairy seeds (e.g., cottonwood). 4. Based on seed colour: Colourless seeds (e.g., *garlic*), white seeds (e.g., *almonds*), yellow seeds (e.g., *daffodils*), green seeds (e.g., beans), *brown seeds (e.g., chestnuts*)	Based on the provided context, the four generally recognized classes of seeds are: 1. Breeder seed 2. Foundation seed 3. Registered seed 4. Certified seed	There are four generally recognized classes of seeds. They are Breeder seed Foundation seed Registered seed Certified seed. The basis of seed multiplication of all notified varieties/hybrids is the Nucleus seed. Definition of seed classes Nuclear seed: This is one hundred percent genetically pure seed with physical purity and produced by the original breeder/institute/State Agriculture University (SAU) from basic nucleus seed stock
What is Certified seed?	Certified seed is seed that has been inspected, tested, and certified by a recognized seed certification agency to ensure its quality and purity. The certification process involves testing the seed for various factors such as germination rate, purity, and freedom from disease and pests. The main benefits of using certified seed include: 1. Improved crop yields: Certified seed is more likely to result in higher crop yields due to its improved genetic quality and greater germination rates 2. Increased profitability: By using high-quality seed, farmers can increase their profits by producing higher yields at harvest time 3. Reduced risk of disease and pests	Based on the provided context, here's a brief summary of what certified seed is: Certified seed refers to the progeny of foundation seed produced by registered seed growers under the supervision of seed certification agencies. The goal of producing certified seed is to maintain the seed quality as per minimum seed certification standards. A blue certificate is issued by the seed certification agency for this category of seed. In other words, certified seed has been tested and inspected to ensure that it meets certain standards of quality, purity, and genetic identity. This is important for farmers, seed sellers, and consumers because it helps	Certified seed: The progeny of foundation seed produced by registered seed growers under supervision of seed certification agencies to maintain the seed quality as per minimum seed certification standards. A blue colour certificate is used by seed certification agencies for this category of seed. Labelled Seed: The seed notified under Section 5 of the Seeds Act, 1966, such seed sold in the market has to be labelled as prescribed under Section 6(a) and (b) of the Seeds Act Such seed is called Labelled Seed. The foundation and certified seeds can be multiplied at stage 1 and II, but the reproduction cannot exceed three generations after breeder seed

(*continued*)

TABLE 2.1 (Continued)

Query	LLM response	RAG response	Contextual response
What is the role of Primary Agricultural Credit Societies (PACS) in BBSSL	Primary Agricultural Credit Societies (PACS) play a crucial role in the implementation of BBSSL (Bharatiya Beej Sahayata Limited) by facilitating the production, testing, certification, and distribution of certified seeds. PACS act as intermediaries between farmers and government schemes, leveraging policies and financial assistance to ensure the availability of high-quality seeds. They also support farmers in adopting modern agricultural practices, improving productivity, and ensuring food security. Through PACS, BBSSL aims to strengthen the agricultural supply chain and promote sustainable farming practices	The system is implemented through Primary Agricultural Credit Societies (PACS) by leveraging various schemes and policies of different ministries of the Government of India	Ensuring the role of farmers in production of certified seeds. This society will focus on production, testing, certification, procurement, processing, storage, branding, labelling and packaging of two generations of seeds i.e. foundation and certified (breeder seeds will be sourced from Public Sector Research Organizations and International Research Institutes i.e., ICRISAT, IRRI, CIMMYT etc.) through Primary Agricultural Credit Societies (PACS) by leveraging various schemes and policies of different ministries of Government of India. The BBSSL is promoted by Indian Farmers Fertiliser Cooperative Limited (IFFCQ), Krishak Bharati

2.6 EVALUATION METRICS OF RAG

The performance of retrieval-augmented generation systems is measured by evaluating how effectively the retrieval component identifies and ranks relevant documents or information. RAG evaluation measures the accuracy of your retrieval sentence by counting metrics on top of the results the AGRO-LLM system that will allow you to monitor programmatically your pipeline's accuracy, recall power, and adherence to facts. These metrics ensure that the retriever fetches an appropriate amount of relevant information and presents it in the correct order. The evaluation results obtained using BERT Score metrics are summarized in Table 2.2.

BERT Score evaluates the similarity between predicted and reference texts using contextual embedding's from a pre-trained BERT model. It calculates Recall, Precision, and F1-score based on the cosine similarity of token embedding's as shown in Equations (2.1), (2.2), and (2.3).

$$\text{Precision} = \frac{1}{|P|} \sum_{P_i \in P} \max_{r_i \in P} \text{sim}\left(p_j, r_j\right) \tag{2.1}$$

$$\text{Recall} = \frac{1}{|R|} \sum_{P_i \in P} \max_{p_i \in P} \text{sim}\left(r_j, p_j\right) \tag{2.2}$$

$$\text{F1} = 2\left(\frac{\text{Precision} \cdot \text{Recall}}{\text{Precision} + \text{Recall}}\right) \tag{2.3}$$

where,
$P = \{p_{1,}\ p_{2,\dots\dots\dots,}\ p_m\}$ be the set of predicted tokens (embedding)
$R = \{r_{1\ ,}\ r_{2,}\dots,r_n\}$ be the set of reference tokens (embedding)
sim $(p_j, r_j) = \cos(p_j, r_j)$ be the cosine similarity between embedding

TABLE 2.2
The Evaluation Metrics for RAG Responses, Including Precision, Recall, and F1 Score, based on BERT Score Evaluation

		RAG BERT Score			LLM BERT Score		
	Query	Precision	F1-score	Recall	Precision	F1-score	Recall
1	How do beetles kill the *Parthenium*?	**0.912**	**0.908**	**0.904**	0.859	0.855	0.850
2	What are the benefits of brown manuring?	**0.908**	**0.915**	**0.922**	0.873	0.872	0.871
3	Classify types of seeds	**0.889**	**0.863**	**0.838**	0.778	0.796	0.815
4	What is Certified seed?	**0.888**	**0.877**	**0.866**	0.850	0.846	0.841
5	What is the role of Primary Agricultural Credit Societies (PACS) in BBSSL	**0.931**	**0.873**	**0.823**	0.864	0.851	0.839

Based on the BERT Score analysis, the RAG model demonstrates greater precision and consistency, particularly in fact heavy- and definition-based queries. In contrast, responses generated solely by the LLM are reasonably accurate but tend to be less reliable for classification or process-oriented questions. This suggests that the retrieval component in RAG significantly enhances the quality and completeness of responses, especially in domains requiring structured or factual knowledge.

2.7 CONCLUSION

The global advancement and expansion of agricultural products, along with the potential applications of information technologies especially large LLM's to enhance agricultural practices, have been thoroughly examined. It aimed to set up a baseline for assessing the capabilities of LLM model Lama 3.2 in addressing complex problems in agriculture. This study presented the RAG-LLM framework, an agriculture-specific framework designed to enhance agricultural practical applicability and decision support through the integration of LLMs. Through a comprehensive pipeline involving data collection, semantic embedding creation, and RAG, the RAG-LLM framework serves as a dedicated knowledge base tailored to the unique requirements of the agriculture sector. By evaluating its performance in this context, the study offers meaningful insights into both the constraints and strength of LLM within the agricultural applications. RAG enhances accuracy in large language models by retrieving contextually relevant information, particularly in tasks such as analyzing farm data. The low setup cost of developing embedding, as vectorised representations of data are, makes RAG an attractive one. However, increase in the input token size can make the prompt larger and the output token size more complex and harder to manage. This is a significant contribution of RAG to precision agriculture with a novel approach towards getting the correct and relevant solutions from the agricultural dataset utilizing the RAG method over sending directly to the LLM method.

The AgroBot framework successfully demonstrated the integration of FAISS and Llama 3.2 models to create a robust, context-aware question-answering system. By combining efficient retrieval through FAISS and natural language generation using the Llama model, the system provides highly relevant and accurate answers to user queries. The hybrid architecture ensures efficiency, contextual relevance, and scalability. This study presented an agriculture-specific framework designed to enhance agricultural practical applicability and decision support through the integration of LLMs. The evaluation of RAG responses using BERT Score metrics reveals that the average precision, recall, and F1-score are 90%, 86%, and 87%, respectively. These scores are consistently higher compared to the corresponding BERT Score values obtained from LLM-only responses, showing the capability of the RAG framework to produce more precise and context-aware responses.

REFERENCES

Balaguer, Angels, Vinamra Benara, Renato Luiz de Freitas Cunha, Todd Hendry, Daniel Holstein, Jennifer Marsman, Nick Mecklenburg et al. "RAG vs Fine-Tuning: Pipelines, Tradeoffs, and a Case Study on Agriculture." arXiv preprint arXiv:2401.08406 (2024).

Balpande, M., K. Mahajan, J. Bhandarkar, G. Borse, and S. Badjate. "AI Powered Agriculture Optimization Chatbot Using RAG and GenAI." In *2024 IEEE Silchar Subsection Conference (SILCON 2024)*, Agartala, India, pp. 1–6. 2024, DOI: 10.1109/SILCON63976.2024.10910462.

Banerjee, Saikat et al. "A Study of the Application Domain of a Large Language Models in the Agricultural Sector." *International Journal of Innovative Research in Computer Science and Technology (IJIRCST)* 12, no. 5 (September 2024): 2347–5552, ISSN (Online).

Biplov, Paneru et al. "Leveraging AI in Ayurvedic Agriculture: A RAG Chatbot for Comprehensive Medicinal Plant Insights Using Hybrid Deep Learning Approaches."*Telematics and Informatics Reports* 16 (2024): 100181.

Chia, H., A. I. Oliveira, and P. Azevedo. "Implementation of an Intelligent Virtual Assistant Based on LLM Models for Irrigation Optimization." In *2024 8th International Young Engineers Forum on Electrical and Computer Engineering (YEF-ECE), IEEE,* 2024.Dong, Xiaoning. "AI and Information Systems in E-Commerce: Building Complex Adaptive Business Systems." Master's thesis, California State University, Los Angeles, 2024.

Jeya, R., Dimpy Singh, Anuj Jha, and Soumik Mallick. "Agriconnect: A Data-Driven Platform for Optimizing Crop Production with Fine-Tuned Large Language Models." In *2024 10th International Conference on Electrical Energy Systems (ICEES)*, pp. 1–7. IEEE, 2024.

Kästing, Marvin et al. "Assessing Large Language Models in the Agricultural Sector: A Comprehensive Analysis Utilizing a Novel SyntheticBenchmark Dataset." In *INFORMATIK 2024* (pp. 1279–1286). Gesellschaft für Informatik eV, 2024.

Lewis, P., E. Perez, A. Piktus, F. Petroni, V. Karpukhin, N. Goyal, H. Küttler, M. Lewis, W. Yih, T. Rocktäschel, S. Riedel, and D. Kiela. "Retrieval-Augmented Generation for Knowledge-Intensive NLP Tasks." *Advances in Neural Information Processing Systems (NeurIPS)* 33 (2020).

Muhammad, A. et al. "A Survey on Retrieval-Augmented Generation with Large Language Models." in *Proc. 28th International Conference on Knowledge-Based and Intelligent Information & Engineering Systems (KES), Procedia Computer Science*, Elsevier, 2024.

Nori, H., N. King, S. M. McKinney, D. Carignan, and E. Horvitz. "Capabilities of GPT-4 on Medical Challenge Problems," arXiv preprint arXiv:2303.13375, (2023).

Nunes, L., B. Silva, R. Estevão, V. Aski, and R. Chandra. "Evaluating GPT-3.5 and GPT-4 on Brazil's National High School Exam (ENEM)." arXiv preprint arXiv:2308.03219, (2023).

Parthiban, Eden Samuel, and Mohd Adil. "Trends in the AI-based Banking Conversational Agents Literature: A Bibliometric Review." *Asia Pacific Journal of Information Systems* 33, no. 3 (2023): 702–736.

Samuel, Dinesh Jackson, Inna Skarga-Bandurova, David Sikolia, and Muhammad Awais. "AgroLLM: Connecting Farmers and Agricultural Practices through Large Language Models for Enhanced Knowledge Transfer and Practical Application." arXiv preprint arXiv:2503.04788. (2025).

Selva Kumar S et al. "Overcoming LLM Challenges Using RAG-DrivenPrecision in Coffee Leaf Disease Remediation." *2024 International Conference on Emerging Technologies in Computer Science for Interdisciplinary Applications (ICETCS)*. IEEE, 2024. DOI: 10.1109/ICETCS61022.2024.10543859. 979-8-3503-7250-2/24/$31.00 ©2024.

Silva, Bruno, Leonardo Nunes, Roberto Estevão, Vijay Aski, and Ranveer Chandra. "GPT-4 as an Agronomist Assistant? Answering Agriculture Exams using Large Language Models." arXiv preprint arXiv:2310.06225 (2023).

Singhal, K. et al. "Towards Expert-Level Medical Question Answering with Large Language Models." arXiv preprint arXiv:2305.09617, (2023).

Vizniuk, Artem, Grygorii Diachenko, Ivan Laktionov, Agnieszka Siwocha, Min Xiao, and Jacek Smoląg. "A Comprehensive Survey of Retrieval-Augmented Large Language Models for Decision Making in Agriculture: Unsolved Problems and Research Opportunities." *Journal of Artificial Intelligence and Soft Computing Research* 15, no. 2 (2024): 115–146.

Wei, J. Wang, X. Schuurmans, D. Bosma, M. Ichter, B. Xia, F. Chi, E. Le, Q. and Zhou D., "Chain-of-Thought Prompting Elicits Reasoning in Large Language Models," in *Advances in Neural Information Processing Systems (NeurIPS)*, 2022.

3 A Review on Smart Sensors for Precision Farming in Agriculture Using IoT

Rakshitha D Shastry, Richa Sinha, and Chethana H.T.

3.1 INTRODUCTION

Conventional agriculture practices usually involve manual monitoring and decision-making, leading to inefficient water usage, fertilization, and pest control systems. The rapid population growth alongside climate uncertainties is projected to increase global food consumption in the upcoming years. Precision agriculture has provided a cornerstone solution to increase agricultural efficiency with great effect on minimizing environmental damage [1].

This review discusses upon precision agriculture as an emerging scientific field that utilizes data-driven techniques to improve agricultural productivity and improves farmers awareness about some of the best cultivation practices that will support an increase in agricultural productivity to meet the increasing food demand [2]. Challenges such as high costs associated with technology adoption in precision farming and the lack of technical expertise in small-scale farmers also hinder the implementation of precision farming. The global agricultural sector is undergoing significant transformations. Traditional farming practices, which were mainly based on manual labor and decision-making based on experience, are gradually accepting precision agricultural approaches that rely more on data. As illustrated in Figure 3.1, a precision agriculture system utilizes IoT sensors, drones, and AI for real-time monitoring and resource optimization [3]. The innovation in the domain of IoT and big data analytics makes precision agriculture possible as the technologies help farmers monitor and manage agricultural activities, primarily, irrigation, fertilizers, and pests in real time [4]. Machine learning algorithms such as decision trees, random forests, and neural networks augmented by smart sensors assist in expressing and processing large-scale agricultural data leading to increased efficiency and lowered environmental impact.

In this study, the revolutionary power of IoT technologies in the agricultural domain is reviewed enhancing the sense of human–machine interaction and the successful utilization of real-time information have been stressed. Accuracy in farming is facilitated by the use of data collection via the applications such as GPS devices, smartphones, and geospatial data-gathering instruments. The analysis of data is found to be of great importance to predict, control, and help in managing and supporting agriculture. The research differentiates the data security protection of data to that of data privacy-ensuring responsible access to data that are a significant risk to the privacy of smart farming, i.e. data theft and unlawful surveillance. A number of contemporary techniques that protects the agricultural data are encryption, blockchains, and multiple legal approaches.

This chapter is structured as follows. Section 3.2 discuss the existing smart agriculture technologies, and related problems. Section 3.3 explains the pros and cons of IoT in precision agriculture. Section 3.4 explores the future perspective of smart farming, new technologies, and capacity building; and Section 3.5 provides a summary of conclusions and aspects, that make smart farming sustainable in its application.

 DOI: 10.1201/9781003613510-3

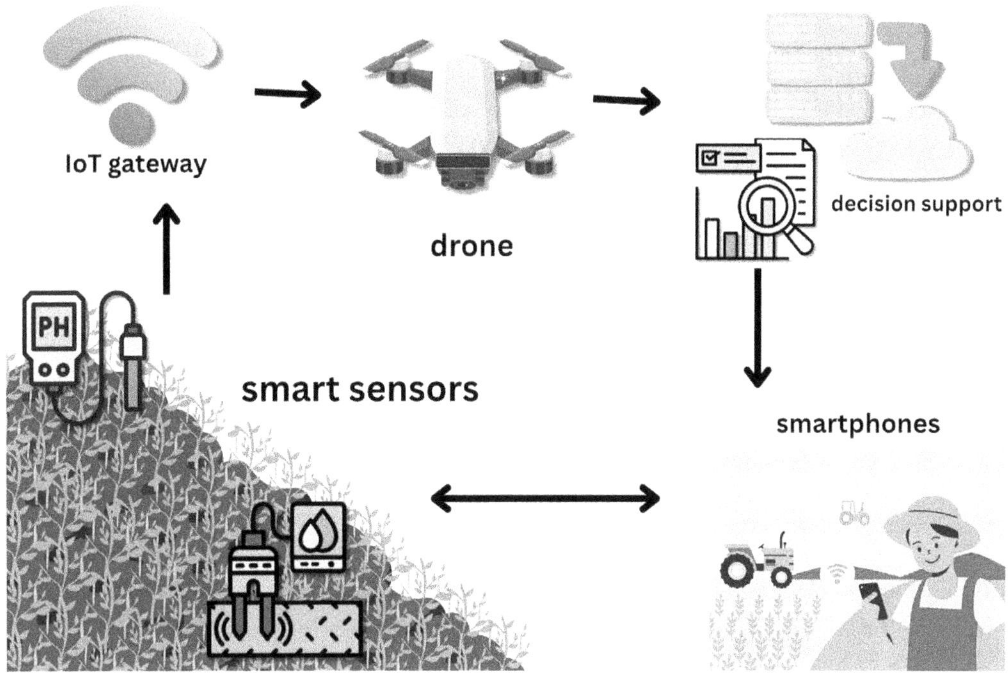

FIGURE 3.1 Precision farming system.

3.2 EXISTING TECHNOLOGIES AND APPLICATIONS

Over the past few years, the smart agricultural systems have gathered a lot of attention due to their potential in ensuring the efficiency and productivity in agriculture. Analysis of this trend have been provided by some of the review papers, highlighting that there exist several ways in which smart agriculture systems may be utilized, namely, precision farming, irrigation management, and crop health analysis. These papers outline the different ideas and designs that include measurements of the environmental factors (including temperature, humidity, and soil moisture), automated actuators, module of communication, or cloud application to make real-time decisions. Through these technologies, smart agriculture systems have already been implemented in various forms of crop farming that has led to improved use of resources, fewer labor requirement, and establishment of an efficient method of farming.

The paper, entitled "Precision Agriculture Using IoT Data Analytics and Machine Learning", is authored by Akhter and Sofi et al. [1]. In this study, it has been observed that crop diseases, pest management, and yield monitoring are critical matters in precision agriculture. The authors note that the crop diseases including the leaf spots, the apple scab, and the late blight are a serious economic risk to agriculture as they have a loss potential worth billions of dollars. The FAO estimates that crops are affected by diseases and pests leading to about 20–40% crop losses annually followed by another bottleneck that is addressing the problem of crop loss. To counteract these agricultural threats, wireless sensors, drone services, the Internet of Things, and computerized knowledge synthesis can be perceived by an overall follow-up on plant health and some preventive measures must be taken early to control the disease spread, using proactive methods. Although pest management is proactive, in this case, there is yet again an argument strong in support of integrated pest management. IoT devices can also forecast the attacks of the pests by measuring the parameters of environment and soil and enabling the farmers to intervene with the pests before they strike. This saves the use of chemical treatments and health dangers to fauna. Another point that the authors make concerns the spatial implementation of IoT sensors that identify pest locations and environmental factors allowing the management of pests more efficiently and in a controlled manner [5]. In the case of yield monitoring, the paper shows the implementation of the IoT technologies to monitor crop performance over the period of planting to harvesting. These consists of the data gathered by various optical sensors and other IoT devices to guide farmers on the status of the crops that they should take necessary actions to store, transport, and cure the diseases during the harvesting process.

Among the several solutions that have been suggested to combat these challenges, there have been development of various complex algorithms to enhance resource allocation and ensure productivity maximization. Additionally, development of robust sensor networks using mesh topology can effectively collect data, and when combined with machine learning and IoT, they can boost disease prediction and improve the crop quality. The necessity of physical protection of the IoT devices against harsh environmental conditions such as rain and wind has been highlighted along with the importance of installation of backup systems to guarantee continuity and dependability. The study also presents notable results, including the creation of an apple scab disease prediction model based on IoT and machine learning. Key findings include the variations in incubation of apple scab across different locations, identification of certain zones as unsafe for cultivation and better disease prediction and control compared to traditional methods.

The key suggestions identified to overcome the challenges are the evolution of sophisticated algorithms to optimize resource allocation, establishment of powerful sensor networks with mesh topology to facilitate data gathering, integration of machine learning and IoT data analytics to optimized disease prediction, and works on crop quality. Moreover, this paper indicates the necessity of physical security to insulate the IoT mechanisms against climatic variations like rain and winds, powering up the backup systems to ensure uninterrupted reliability and operating support. Other prime results contained in this rating include the development of a prediction model for apple scab infection using IoT and machine learning keywords for variable incubation periods of apple scab disease in different regions, naming certain areas unsafe for cultivation and increased efficiency in disease prediction and response time compared to existing conventional practices.

Further a conclusion can be drawn that IoT and machine learning, when infused with precision agriculture, will radically change crop disease prediction and management. That said, also an issue is raised of overly high initial costs, training the involved personnel, and various environmental factors involved. Thus, the authors promote the idea of smart agriculture practices on the premise that overcoming these barriers will yield long-term benefits for crop management and production.

Another paper titled "Precision Agriculture in Modern Farming" by Fantin et al. [6] summarizes certain critical challenges related to agriculture and the precise technologies implemented to resolve them. Crop diseases presented a major challenge to farmers that need wide experience and background for diagnosis. To overcome this, the authors proposed that computerized systems for real-time pest and disease detection can be developed using modern algorithms. Data are analyzed and control measures prescribed to ensure more accurate disease management and minimized yield losses.

The authors emphasize using precision farming techniques in input management that enable farmers to apply inputs depending on to specific needs. This involves the applications of AI and IoT to track environmental condition and crop health monitoring so that informed decisions about the use of fertilizer and pesticides can be taken. For instance, Contamination can be prevented during the selection of seeds which are resistant to local diseases there by reducing use of fertilizers.

Another area of concern is the precision with the employment of resources, where the conventional farming has hardly been given the opportunity to have accuracy required to reduce resource utilization. The authors advocate the use of precision agriculture technologies that can measure, examine, and analyze the needs of individual fields and crops so that the crop production is optimized. This would benefit small family farms and cooperatives who may lack the ability to make high IT investments. Monitoring and controlling play a significant role in agricultural management.

Here, AI-based food monitoring and quality control systems help in improving grading, storage, and drying of harvested crops which are produced, representing the attributes of certain quality in them thus drawing the attention of the author. These systems provide real-time data and insights that can minimize the losses after harvest assuring the crops meet the market standards.

On the other hand, the authors state that while equitable access to the technologies is most pronounced in larger operations that will benefit from the technology because they must make an investment for that. This does not apply for technology use by smaller ones. This problem can be resolved through increased accessibility for smaller farms and collaboration and sharing of resources among cooperatives and small family farms.

Finally, Fantin et al. [6] consider the environment to be an important part of agricultural practices, and hence, the paper highlights the need to manage spatial and temporal variability in production. Precision agricultural practises can significantly improve the performance of the crops in terms of both quality and quantity. Thus, it is an opportunity for precision farming to make modern farming more productive and sustainable.

Building on these concepts, yet another paper by Geetha Chellaian et al. [7] proposed an AGRIOT model that could improve the agricultural productivity with sensors and IoT by real-time monitoring and implementation of data-driven decisions. Their strategy lies in maximizing the resource utilization and thus improving farming efficiency by utilizing advanced technologies. Some of the issues in traditional farming involves inefficiency in resource utilization, environmental deterioration, and need for larger food production. Conventional agriculture leads to excessive use of fertilizer and pesticides, soil erosion, and water wastage. Smart farming is the process of combining IoT and smart sensors to systematically streamline all the individual processes providing numerous management methods ensuring sustainability and productivity.

A smart farming system is a combination of multiple components that are effectively working together to enhance agricultural productivity. Soil management systems use sensors which can monitor nutrient levels, moisture content, and pH of soil, therefore, assisting farmer's decision on soil amendments and crop rotation methods to improve soil fertility and yield. The selection of crops involves the utilization of data analytics and predictive models to evaluate the historical and the current environment, to choose the type of crops that should be planted in various soils and weather conditions. The smart farming systems analyze the nutrient levels of the plant and suggest the fertilizers that would help them grow without overuse or incurring excessive costs in the process. Insect and weed control operate through sophisticated imaging and sensor technologies to detect pests and weeds in real time and either automate or direct interventions with minimum application of chemicals. Water management/irrigation automatically adjusts irrigation through IoT devices that respond to soil moisture and weather conditions to ensure the most efficient water application and prevent wastage. Weather prediction utilizes forecasting models to assist farmers in preparing for climate change and changing their farming strategies based on these predictions. Crop yield prediction utilizes data from sensors and weather predictions to predict yields accurately, allowing for prior planning and allocation of resources. Marketing the products is when smart farming systems come in to assist market analysis for farmers regarding the best time to sell their products based on predicted yields and demand trends.

All sensors are placed every 100 feet in a 1-acre banana farm and connected to an Arduino board to collect sufficient data. The sensors continuously collect the data on soil conditions, weather, and crop health [5]. These data are sent to a central system for analysis [8]. Training of ANN becomes significant as different ANN models must be appropriately trained to accurately predict yields for various crops based on environmental factors. The iterative process allows the system to detect errors, make predictions, and continue to increase the accuracy of predictions compared to actual outcomes, retraining with current data for system reliability. Insight enabled through data would allow farmers to address such issues as nutrient deficiency, pest infestation, and irrigation needs. This ANN model for yield prediction recognizes input, hidden, and output layers that adjust accuracy through continuous learning with the use of a sigmoid activation function for better predictions. The iterative process means the system can adjust to newly sprouting conditions agriculture can discover through IoT and AI to increase productivity, minimize wastage of resources, and allow farming to be sustainable. To conclude, the usage of the IoT and sensor technologies integration into smart farming provides one of the solutions to many challenges which modern agriculture faces. Such a system improves every farmer's means to be resource efficient in the decision-making process and promotes sustainable farming by increasing crop yield. Overall, the proposed methodology provides an insight about implementing this novel approach in agricultural settings.

Building on the theme of smart agriculture, yet another approach was described and proposed by Gaikwad et al. [9], the said approach is an Arduino-based IoT system which uses low-cost sensors for real-time monitoring of soil and air. This system architecture adopts sensors and the sensor web enablement (SWE) architecture, consisting three main components: an Arduino kit with sensors, an Android application [9], and a cloud-based database server. Its connectivity with the mobile device is made possible by the Wi-Fi microchip ESP8266, thus facilitating portability and making it cost-effective. Even though Bluetooth and GSM modules are on the list of choices, it is the Wi-Fi that has been adopted for being both effective and economically viable at the same time, having larger transmission range. The IoT kit was developed initially to monitor crop health in controlled environments; hence, it has now been included with other sensors such as those monitoring air and soil temperature, humidity, and soil moisture.

The RC-A-4079 moisture sensor, DS18B20 soil temperature probe, DHT11 air temperature and humidity sensor, a source of power, and an ESP8266 Wi-Fi module are included in the IoT kit. The circuit is enclosed in a plastic container, with data transmission enabled through a mobile hotspot. The DHT11 sensor measures humidity levels between 20% and 90% with ±5% accuracy and temperature between 0°C and 50°C with ±2°C accuracy. The Arduino board, programmed using C++, was designed with a steel probe and an Android application to measure soil moisture at the root zone level [9]. The Android mobile application is developed using Android Studio version 3.1. The application captures location information (spatial data), identifies different soil types, and determines the soil moisture levels. It uses the smartphone's GPS capabilities and the Volley library for Internet communication as well as the JSON file format for transmitting data back and forth to/from the server. A web-based database server is created using PHP, MySQL, and the Leaflet GIS mapping library.

An adjustable, detachable soil sampling probe is created by adapting a 40 cm long steel pipe into 10 cm long sections depending on the crop being studied or the soil conditions. Studies have shown that the root depth varies by crop, thus the need for accurate soil moisture measurements to optimize agricultural production.

The experimental system uses a structured registration process; user registration, setting up sensors which included a unique serial number for each sensor and establishing project configuration. The environmental parameters are continuously sent to a cloud server via the IoT kit. Validation of these continuous readings were completed during a pilot study conducted in agricultural fields on groundnut and wheat crops at several different depths.

The Android Soil Mapping App provided the ability to select a project, record GPS locations, acquire sensor readings, and temporarily store the data online so users could view their collected data using their smartphones. Real-time viewing of the recorded data was also possible for users to see using the user dashboard that displayed mapping tools, GPS locations, timestamps, and data markers.

An experimental Assessment was conducted using the sensor data from the system versus a commercially available Vantage Pro2 weather station located at JNE College. Both the VH400 soil moisture sensor and the DS18B20 temperature sensor were reliable, with the DS18B20 demonstrating a low root mean square error (RMSE) of ±1.06° C for a 10-day experiment. The overall accuracy of the system was reflected in RMSE values of ±3.97% for soil moisture, ±0.7° C for air temperature, and ±2.14 % for air humidity.

In comparison to other IoT solutions, this application costs less than many alternatives, and therefore, it provides a reasonable trade-off between cost, accuracy, and functionality. The app consists of a data logger, a digital display, and wireless communication, making it suitable for use in many agricultural applications.

This innovative approach helps farmers with timely data to support their decisions on the use of water and to avoid crop failures. Advanced sensor technology allows practice of precision agriculture and combines sustainable food security with stewardship of the environment in farming communities.

Following these, Amiri-Zarandi et al. [10] emphasize that substantial quantities of data are generated and processed within smart farming practices. Hence, there are painful concerns over security and privacy. Data security excludes data from attacks on them, while data privacy restricts the collection and access to data according to users' preferences. Privacy, though normally stated, restricts unauthorized access and disclosure, which often differs in its application across jurisdictions. Where IoT-based farming is concerned, security concerns have dominated the surveys of prior research, while privacy problems have only had limited focus. There are few papers which addresses security and privacy challenges in IoT-based farming. For instance, an existing review [10] has explored security threats in smart farming environments, including IoT vulnerabilities and cyber-physical system risks. However, these studies mainly emphasize security threats rather than privacy solutions for the said threads, urging the need for further investigation into privacy-preserving mechanisms. Furthermore, this paper recognizes many privacy-enhancing technologies relevant to smart farming and explores various technological applications in this field as discussed below.

- **Machine Learning**
  Machine learning is the technology where machines are trained and they learn from experience. This way it helps farmers to analyze data and get improved decision-making and business performance. Farm data are collected from smart sensors, which machine learning models will process and those help for yield prediction, water management, and disease detection. Privacy is crucial, and studies have developed solutions to protect sensitive information while using machine learning in agriculture. Using the concept of Machine Learning, a model was created to detect suspicious sensors in forest environments by studying how environmental factors influence their behavior. Similarly, a deep learning system detects diseases in apple trees using images, allowing data use without sharing. A framework for weed quality analysis employs noise to protect data privacy while identifying low-quality products. The IoT use cases take place in machine learning for device authentications and anomaly detection which improves smart farming security. IoT environments demonstrate intelligent access control mechanisms using machine learning to discover and control unauthorized processes. In federated learning, the model uses datasets that exist in different devices or are different datasets (customers, individuals) with data remaining on the local devices preserving data privacy.
- **Edge Computing**
  In smart farming, two types of communication are used: device to device (through direct connection) and device to server (through cloud servers). A recent solution uses edge nodes as middleware for near data processing before transmission to maintain data privacy and reducing data volume. The study presented in use cases is an example of how edge computing decreases the need for transferring data from one location or farm management to another, and this is also used in pest control. Edge-based farming considerations include and analyze

such aspects as device location and consumption simulation through simulators. Using AI, decision support systems predict crop yields and droughts and also lower the amount of data that must go from the devices to the main servers. Low-cost edge server, such as Raspberry Pi assist in monitoring the animal welfare by controlling environmental factors. Generally, edge computing increases security in smart farming while decreasing demand on the cloud servers and improving data traffic.

- **Blockchain**
  Blockchain is new technology that records transactions on a public ledger, with original design based on Bitcoin, and now being implemented in various sectors like data management and smart contracts. The technology improves privacy as it removes the process of central authority and all users have a copy of the ledger, making it transparent and unchangable. Some of the vital characteristics are permission structures for checking a person's identification, consensus approaches to insert new information, and smart agreements for automation. This technology is improving the food traceability, supply chain, and water management in the agriculture where farmers save their data safely on the blockchain, gaining investments. Smart contracts simplify fish farming requirements by removing the interaction with the third parties and reducing the amount of confidentiality breaches. Following the same concept, a blockchain-based mobile app could help Indian farmers track, monitor, and manage agricultural activities, improving increasing trust in IoT-based data [11].

To conclude, privacy is an important component in data integrity and confidentiality. Another standard technique in data protection during the lifecycle of smart farming is evolving. Standardization helps to address the problem of IoT devices and technologies diversification to enhance privacy solutions. To simplify data sharing between the people involved in smart farming, trust is required and hence, trustworthiness measurement can result in positive interaction between farming parties. Legal frameworks play a crucial role in outlining the liabilities and is the mandatory solution to call to account on data privacy in smart farming. The blockchain technology seems to be effective, but even the problems of scaling and overconsumption of general energy seem to contradict all the real-life solutions. Agriculture has been transformed by Technology(s), with the aspect that farmers do not want to share data making the privacy issue even more serious. Privacy at every data lifecycle stage makes the data sharing and smart farming practices more advanced. Smart agricultural technologies have the potential to bring big returns to the agricultural production in the world. The review concludes that there are a lot of remedies that may be used in addressing the frequently paralyzing problem of privacy. They can be discovered and cultivated, without suffocation of the transition to new technologies and large datasets in the agricultural ecosystem.

3.3 ADVANTAGES AND DISADVANTAGES OF PRECISION FARMING

In this section, various different advantages and disadvantages of precision farming is depicted. Table 3.1 presents some of its characteristics including data collection, resource efficiency, automation, disease detection, energy consumption, and scalability, and it illustrates its strong and weak points [12].

The table provided summarizes the key benefits and drawbacks of precision farming, showing how technological solutions assist in efficiency as well as raise challenges that would have to be overcome. Although use of smart sensors in precision farming provides significant advantages in terms of making decisions, resource usage, and automation, there are few issues with them, and they include high prices [3], high energy consumption, and privacy issues. The adoption and implementation of IoT in the agricultural sector should be balanced in balancing these aspects.

TABLE 3.1
Advantages and Disadvantages of IoT Technologies in Precision Farming

Sl.no.	Aspect	Advantages	Disadvantages
1.	Data collection	Sensor technologies are capable of instantaneously collecting vast amounts of data on parameters like soil moisture, temperature, and crop health and therefore allow precise monitoring and management of agricultural activities [6, 13, 14, 8]	Privacy and security threats might arise due to large amount of data collection and may lead to potential latency issues due to the requirement of data transmission and processing [15]
2.	Resource efficiency	IoT-enabled smart sensors help optimize resource use by providing insights for precise irrigation, fertilization, and pesticide application, reducing waste and improving sustainability [16, 17, 14]	The initial setup and maintenance of IoT systems can be costly, and there may be challenges in integrating these technologies with existing farming practices [1]
3.	Automation	Automation of farming operations, such as irrigation and pest control, is facilitated by smart sensors, reducing manual labour and increasing operational efficiency [14, 18]	Automation may replace employment opportunities in rural areas with the inherent disadvantage of excessive dependence on technology, which is highly detrimental in case of a system failure [14]
4.	Disease detection	Smart sensors can detect diseases early by monitoring environmental conditions and plant health, allowing for timely interventions to prevent crop loss [19]	The accuracy of disease detection can be limited by sensor quality and environmental factors, potentially leading to false positives or negatives [19]
5.	Energy consumption	Advances like TinyML allow for energy-efficient data processing directly on sensors, reducing the need for constant data transmission and conserving battery life [15]	Despite improvements, wireless transmission methods still demand high energy, which can be a limitation for remote or off-grid farms [15]
6.	Scalability	IoT systems can be scaled to cover large agricultural areas, providing comprehensive monitoring and control capabilities across diverse farming environments [13, 18]	Scalability can be hindered by connectivity issues, especially in remote areas where network infrastructure is lacking [13]

3.4 FUTURE SCOPE

This area of using smart sensors & IoT technologies in precision agriculture created space for many new areas of discussion and future research needed to increase both sustainability and productivity in agriculture. The following section highlights a few of these areas:

1. **More Sophisticated AI/ML Algorithms:** Research on developing more sophisticated artificial intelligence (AI) algorithms will allow for the analysis of increasingly large amounts of data generated by precision agriculture, which will result in improved predictive analytics for crop yields, pest outbreaks, and resource allocation. The practice of deep learning techniques to precision agriculture will result in more adaptable and advanced agricultural systems. As noted by Kamilaris and Prenafeta-Boldú et al. [20], AI has been demonstrated to have the potential to alter the decision-making framework of farmers through its application.
2. **Printed Sensor for Smart Farming:** Low cost printed sensor technologies are used to monitor a large range of parameters in agriculture. These sensors can be printed on many

inexpensive, flexible materials and can be applied to many different uses in smart farming such as monitoring soil conditions and microclimates [21].

3. **Interdisciplinary Collaboration for Climate Change:** The future of precision agriculture will benefit significantly from interdisciplinary collaboration between agriculture, environmental science, data analytics, and social sciences, and therefore, the whole process of collaborating is particularly important for developing strategies that are adaptable to the impact of climate change on agriculture.
4. **Development of better in Data Privacy and Security Systems:** With the expanded use of IoT in the agricultural sector, the current smart farming requires better mechanisms of data privacy and security development. Additionally, the development of an appropriate regulatory framework for smart farming and ethical guidelines related to data ownership, access, and the distribution of power among stakeholders is also critical.
5. **Education and Training Programs:** Future directions should come up with educational and training programs whereby education and training should be done. Address the difficulty of small-scale farmers having inadequate technical capacity. Such programs empower farmers to master the ability to implement the smart farming technology and enjoy the consequences entirely. This assistance might be provided through the joining of hands with agricultural extension services or learning institutions. How exactly to enlarge precision agricultural technologies to smallholder farmers will have to follow. in research. The new opportunities can involve financial processes or new community-based training.
6. **Incorporation of New Technologies:** The need of incorporating new technologies should also be factored in the future. Such as, implementing newly developed technologies like drones, robotics, and augmented reality into precision agriculture. Development of these solutions will further increase operational and crop management.

The smart farming with smart sensors has resulted in the future that is on the verge to grow tremendously and with increased pace due to technology and the necessity to have efficient and sustainable farming. However, concurrent implementation of new solutions as well as regulatory services is needed to ensure safety against data privacy. When appropriate measures are taken, smart farming can guarantee optimal productivity in sustainable operations while protecting sensitive information.

3.5 CONCLUSION

To conclude, the application of smart sensors and the Internet of Things (IoT) in precision agriculture is definitely transforming the contemporary farming methods because it can solve issues of resource efficiency, environmental issues, and the growing world food demand. These technologies create a holistic system of monitoring in real time, execution of decisions based on data, as well as the use of resources through optimization which improves the productivity, sustainability, and efficiency of agriculture.

Nevertheless, the cost of implementation remains high, and some small-scale farmers lack technical skills. Also, there is the issue of data privacy that impedes the use of big data. These obstacles can only be mitigated by designing secure data management systems, using privacy-sensitive systems and endorsing collaborative models to facilitate equal opportunities and execution.

In addition, research and development is an important aspect that should be sustained so that these innovations are made available and useful to the farming communities and especially the smallholders. The promotion of participatory and resource-sharing networks will be the pathway to the more effective, engaging, and diverse agricultural systems that will eventually lead to the creation of the environmentally friendly and strong future of the global farming.

REFERENCES

1. Akhter, R., & Sofi, S. (2021). Precision agriculture using IoT data analytics and machine learning. *Journal of King Saud University – Computer and Information Sciences*, 34, 5602–5618. https://doi.org/10.1016/j.jksuci.2021.05.013
2. Baddi, Y., Maleh, Y., Alsmadi, I., & Lahby, M. (Eds.). (2025). *Generative AI for Cybersecurity and Privacy* (1st ed.). CRC Press. https://doi.org/10.1201/9781003597476
3. Sethi, V. P (2025). *Greenhouse Engineering -Software-driven Design and Modelling for Optimal Microclimate Control.* CRC Press.
4. Alahmad, T., Neményi, M., & Nyéki, A. (2023).Applying IoT sensors and big data to improve precision crop production: A review. *Agronomy*, 13(10), 2603. https://doi.org/10.3390/agronomy13102603
5. Balasubramanian, S., Natarajan, G., & Chelliah, P. R. (Eds.). (2024). *Intelligent Robots and Drones for Precision Agriculture.* Springer Science and BusinessMedia LLC.
6. Fantin, E., Raj, I., Appadurai, M., Athiappan, K., & Raj, E. F. I. (n.d.). Precision farming in modern agriculture. In: Choudhury, A., Biswas, A., Singh, T.P., Ghosh, S.K. (Eds). *Smart Agriculture Automation Using Advanced Technologies. Transactions on Computer Systems and Networks.* Springer, Singapore. https://doi.org/10.1007/978-981-16-6124-2_4
7. Chellaian, Geetha & College, & Rosenthaliana, Studia. (2022). AGRIOT: Precision agriculture using IoT and sensors. *ResearchGate.* Retrieved from ResearchGate.
8. Chouhan, S.S., Singh, U.P., Saxena, A., Tiwari, A., & Jain, S. (Eds.). (2025). *Artificial Intelligence and Computer Vision for Ecological Informatics* (1st ed.). CRC Press. https://doi.org/10.1201/9781003518778
9. Gaikwad, S. V., Vibhute, A. D., Kale, K. V., & Mehrotra, S. C. (2021). An innovative IoT-based system for precision farming. *Computers and Electronics in Agriculture,* 187, 106291. Elsevier. https://doi.org/10.1016/j.compag.2021.106291
10. Amiri-Zarandi, M., Dara, R. A., Duncan, E., & Fraser, E. D. G. (2022). Big data privacy in smart farming: A review. *Sustainability*, 14(15), 9120. MDPI. https://doi.org/10.3390/su14159120
11. Gyamfi, E., Kropczynski, J., Johnson, J., & Yakubu, M. (2024). Internet of Things security and data privacy concerns in smart farming. *2024 IEEE World AI IoT Congress (AIIoT)*, Seattle, WA, USA, 2024, 0575–0583. https://doi.org/10.1109/AIIoT61789.2024.10578972
12. Sumithra, M. G., Sathyamoorthy, M., Manikandan, M., Dhanaraj, R. K., & Ouaissa, M. (Eds.). (2024). *Computational Intelligence in Internet of Agricultural Things*. Springer Science and Business Media LLCPublication. https://doi.org/10.1016/j.indic.2025.100607
13. Alumfareh, M., Humayun, M., Ahmad, Z., & Khan, A. (2024). An intelligent LoRaWAN-based IoT device for monitoring and control solutions in smart farming through anomaly detection integrated with unsupervised machine learning. *IEEE Access,* 12, 119072–119086. https://doi.org/10.1109/ACCESS.2024.3450587
14. Sodhi, G., & Jamwal, P. (2024). Smart farming: Harnessing the power of IoT for agricultural transformation. *2024 5th International Conference on Recent Trends in Computer Science and Technology (ICRTCST)*, 406–412. https://doi.org/10.1109/ICRTCST61793.2024.10578369
15. Chollet, N., Bouchemal, N., & Ramdane-Cherif, A. (2022). TinyML smart sensor for energy saving in Internet of Things precision agriculture platform. *2022 Thirteenth International Conference on Ubiquitous and Future Networks (ICUFN),* 256–259. https://doi.org/10.1109/icufn55119.2022.9829675
16. Pyingkodi, M., Thenmozhi, K., Nanthini, K., Karthikeyan, M., Palarimath, S., Erajavignesh, V., & Kumar, G. (2022). Sensor based smart agriculture with IoT technologies: A review. *2022 International Conference on Computer Communication and Informatics (ICCCI),* 1–7. https://doi.org/10.1109/ICCCI54379.2022.9741001
17. Soussi, A., Zero, E., Sacile, R., Trinchero, D., & Fossa, M. (2024). Smart sensors and smart data for precision agriculture: A review. *Sensors (Basel, Switzerland)*, 24(8), 2647. https://doi.org/10.3390/s24082647
18. Almalki, F., Soufiene, B., Alsamhi, S., & Sakli, H. (2021). A low-cost platform for environmental smart farming monitoring system based on IoT and UAVs. *Sustainability,* 13, 5908. https://doi.org/10.3390/su13115908

19. Durai, R., Vijayakumar, R., Lakshmisridevi, S., Bhanu, S., & Arunkumar, U. (2024). IoT based optical sensor network for precision agriculture. *2024 International Conference on Optimization Computing and Wireless Communication (ICOCWC)*, 1–7. https://doi.org/10.1109/ICOCWC60930.2024.10470879
20. Kamilaris, A., & Prenafeta-Boldú, F. X. (2018). Deep learning in agriculture: A review. *Agricultural Systems,* 187, 103043. https://doi.org/10.1016/j.compag.2018.02.016
21. Rayhana, R., Xiao, G., & Liu, Z. (2021). Printed sensor technologies for monitoring applications in smart farming: A review. *IEEE Transactions on Instrumentation and Measurement,* 70, 1–19. https://doi.org/10.1109/TIM.2021.3112234

4 Revolutionizing Crop Health Management

Harnessing IoT for Precise Leaf Disease Detection in Agriculture

Rajeshwari J, Madhura J, Shalini K B, Bharath B C, and Hong Lin

4.1 INTRODUCTION

This section sets the scene for understanding AgriLink. This chapter introduces a handful of critical concepts, including overview of the work, problem statement, objectives, and motivation for AgriLink. We shall try to create a foundation to get better insight into the work and its impact.

4.1.1 Overview

Agriculture is an important sector of the Indian economy, supporting the livelihoods of a large population and ensuring food security. However, it is confronted with formidable challenges, including soil erosion and changing climate, as well as a lack of access to modern farming technology. The launch of AgriLink is a bold and focused effort aimed at shaping the industry through technology.

AgriLink utilizes computer vision, machine learning, and deep learning to provide data driven insights as well financing for farmers. The company's overall mission is to provide farmers with the tools and knowledge that they need in order to get more for their resources, production process, and long-term farming. AgriLink is going to solve the current problem of Indian farmers, and we will have a precision farming.

At the farm gate, AgriLink solves some common farmer problems such as soil testing, crop choice, and access to finance. Using digital image processing methods AgriLink helps farmers identify their soil types and decide which crops to plant and what fertilizers need to be applied. AgriLink's crop advice recommendations are individually designed, taking account soil types and matched against the databases on weather recordings and learning machine algorithms to ensure maximum yields.

It also helps manage the risk and finances of farmers by providing loan analysis and revenue forecasting through an integrated module of AgriLink. Using geospatial technology for estimation of land extent and market rate AgriLink enables farmers to review their financial capacity as well as avail institutional credit. KisanRaja has another segment, AgriLink, which is a marketplace where farmers can purchase or sell/rent agricultural machinery and products from each other, creating community relationships while supporting the sharing of resources.

In sum, AgriLink is a model for Indian agriculture in the new age where farmers are not rendered receivers of technology but subjects of it; not mere survivors by day, but livelihood doers – and that into future center stage players. Utilizing an integrated approach and solutions, AgriLink will help drive positive change and unleash Indian agriculture's true potential in the digital age.

DOI: 10.1201/9781003613510-4

4.1.2 Key Concerns

The economy of India is based on the agriculture industry, but like all industries, this industry suffers from a number of problems that do not allow it to develop fully and stably. Among these, according to Chakratestimates, a critical one is the shortage of data as well as sources based on artificial technology and machine learning that could be of assistance to farmers in boosting their farming techniques. Conventional agricultural tools lead to low productivity, soil erosion, and poverty. Uke: In order to address this issue AgriLink with the help of digital image processing, deep learning and machine learning are being used as a solution that will provide expert advice as well arrange credit for Indian farmers. As per findings of the study, certain significant sectors in Indian agriculture are just categorized as:

1. Soil Analysis: Conventional soil testing is a slow process and may in turn be inappropriate. Physical properties such as soil classification and nutrient values will be attached to the soil images so that farmers can make well-informed choices of crop choice and how best to manage their nutrients.
2. Crop Choice: Selecting an appropriate crop capable to grow under the existing conditions is the most important for attaining highest yield and benefit. AgriLink will use machine-learning algorithms to recommend optimum crop selection to farmers based on soil and weather conditions as well as trends from previous years.
3. Financing Support: Farmers need access to finance (credit and loans) in order to invest in new farm methods and equipment. AgriLink will utilize geospatial technology for the estimation of market rate and calculation of land area to enable farmers to evaluate their financial feasibility and avail institutional credit better.
4. Community Involvement: A robust farmers' community and sharing of resources can enhance productivity and sustainability in farming. AgriLink also has a marketplace component where farmers can purchase, sell, or lease farm equipment and swap produce so farmers can cooperate and share expertise.

All in all, the goal of AgriLink rests with the benefit of technology in actually shaping up Indian farming, thereby empowering the farmers through knowledge, financing, and other supports to confront challenges and thrive under changes the times throw up.

4.1.3 Objectives

The principal aim of AgriLink is to revolutionalize Indian farming by adopting technology to address critical issues confronting farmers and other stakeholders. To facilitate this vision, the following objectives have been set:

1. Increase Productivity: AgriLink aspires to enhance productivity within the agricultural community through mobilization of innovative tools and insight that will facilitate the analysis of soil, crop, and resources. The other feature of AgriLink is the decision support system, which is based on the use of digital image processing and machine learning to help farmers take data-based decision-making on getting the maximum out of the crops for cost minimization.
2. One of the facets of precision farming is at micro-level fine-tuning practices based on variability in soil features, climate, etc. AgriLink is precision agriculture simplified because it provides recommendations for plantations based on the sensorial nutrients in the soil, soil pH, and weather forecasts. Based on individualized recommendations, farmers pick the best crops they themselves can grow given certain constraints in their fields.

3. Encouraging Financial Inclusion: Farmers need money to buy machinery, inputs, and invest in better methods of farming. AgriLink promotes financial inclusion through credit rating, revenue estimation, and cash flow planning. Farmers goals could be to have access to credit and loans, for better financial stability and resilience. Geospatial analysis tools coupled with ML algorithms will offer them credits from institutional sources.
4. Community Empowerment: The AgriLink marketplace feature supports supporting community empowerment of farmers to sell, buy, or barter goods, i.e. farm produce or machinery. Peer-to-peer trade and exchanging information formulate a system that AgriLink believes will empower farmers to utilize pooled wisdom and resources to enhance their livelihoods and surmount challenges.
5. Enable Sustainability: Sustainable agriculture implies the application of practices that are economically sustainable, socially acceptable, and environmentally friendly. AgriLink enables the efficiency of resource utilization, minimal environmental degradation, and soil fertility and biodiversity-promoting practices for achieving sustainability. AgriLink combines agricultural management with a focus on the long-term sustainability and resilience of Indian agriculture and therefore the livelihoods of the farmer.
6. The objectives of AgriLink are to empower Indian farmers with access to technology-driven solutions for increased productivity, promote financial inclusion, foster community engagement, and advance sustainability in agriculture. Addressing this objective will convert Indian agriculture into modern, resilient, and inclusive sectors that can handily face the challenges of the future.

4.1.4 Motivation

The problem AgriLink seeks to address is the immediate demand for a reimagined and optimized Indian agriculture sector given that it is at the hub of the economy, and yet, it provides subsistence to millions. And as vital as agriculture is, it would seem to have sunk in a cocktail of issues that stand between agriculture and productivity and resilience in the face of changing times (environmentally and economically speaking). Some of these problems include:

1. Productivity and Yield: The production yields that farmers are achieving in India is below optimal. Reasons for this is soil degradation, water shortages, and climate changes among other factors. Conventional methods of farming will not be able to meet the increasing need of food and agricultural products, hence a demand for new technologies/data-based models to achieve high yield while meeting the demand.
2. Resource Management: Soil, water, and fertilizers must be managed efficiently as a matter of necessity for sustainable agriculture. However, they do not have enough knowledge to be able to improve the way resources are managed including inputs and outputs, which leads to environmental degradation due to inefficiency.
3. Financial Accessibility: Credit Access, including loans and credit, has been one of the important constraints to face by smallholder farmers in India; particularly, poor access to institutional credit due to high interest rate hindered the farmer investment in modern farming practices and other equipment as well as inputs, which ultimately restraining their productivity as well income.
4. Market Access: Farmers suffer hurdles regarding the market access to sell their outputs at the best prices. This is in addition to the absence of adequate market information, oppression by middlemen, and flawed infrastructure which makes the farmers weaker and probably losing on income.

Working together with farmers creating actionable insights with whatever resources available, AgriLink plans to develop itself by establishing enterprise-grade digital image processing, machine learning, and geospatial analytic systems. Attractiveness to the development of science-based agriculture, dissemination of the knowledge of soil analysis and crops suitable for that soil, and availability of financial planning sets will go a long way toward raising productivity, profitability, and sustainability.

In addition to building great space for community engagement and sharing of resources through its marketplace feature, AgriLink intends to strengthen and create an agricultural ecosystem that stays strong and supports all farmers throughout the nation. The ultimate aims are to see Indian agriculture blossom as modern, data-led, and inclusive so that it can survive and thrive under the uncertainties of the future.

4.1.5 Analysis

The agriculture sector of India is facing multiple problems of various nature hampering productivity, sustainability, and economics of agriculture. A thorough analysis has led to the identification of some problems that AgriLink will take up for appropriate actions.

1. Technological Access: Most Indian farmers, being denied an opportunity to apply their knowledge for adopting improved agriculture methods and technology, have been deprived of high-yielding agriculture practices and would therefore be unable to increase their productivity or efficiency. Traditional farming relies on old techniques and lacks adequate information, a major cause for low crop yields and wasting precious inputs like soil, water, and chemicals.
2. Land Degradation: Agricultural productivity is plagued by soil and nutrient degradation. If farmers did not conduct soil sampling and do nutrient management correctly, this could result in a situation where crop choices and fertilizer application rates were suitable only with regard to the maintenance of soil health.
3. Variable Markets: Farmers can never be certain of the demand in their markets, price indices, and even having access to only markets. Market intelligence and development of the distributional channels were never established to satisfaction. Inefficiency thus comes as a result of failure to transport farm produce to the market and pulling our farmers into running loss (vicious economic conditions).
4. Access to Finance: Access to finance and credit is critically challenging for many farmers in general, especially smallholders who are located in remote areas. With no collateral and farm credit history, and with only ill-shaped banking services, these farmers are unable to risk investments on agricultural inputs, i.e., intermediate technologies and technology upgrades.
5. Information and Decision Making: Information and Decision Making: The great majority of farming decisions taken by a farmer are based on subjective judgment and little knowledge. Lacking data-driven insights and decision-support systems, a farmer may hardly optimize decisions concerning crop selection, irrigation scheduling, pest management, and resource allocation.

   Limited Community Collaboration: Differentiated and disparate agricultural ecosystems of India are limiting collaboration, knowledge sharing, and pooling of resources in farming. The farmer community does not have access to platforms or market integration tools to leverage collective wisdom, economic dynamics, and resources for mutual benefit.

AgriLink seeks to tackle this issue through the use of digital image processing combined with machine learning and financial appraisal principles in order to assist farmers with actionable insights, method, and services for finance. It aims to improve farmer productivity, profitability, and viability in India by empowering them with technology-driven solutions and fostering community engagement.

4.2 METHODS AND MATERIALS

4.2.1 Hardware Requirements

In order to assure optimum performance and regulatory functioning of AgriLink, the following pieces of equipment will be, therefore, required:

1. AgriLink is designed for operating on Android and iOS smartphones or tablets, and thus, it will require smartphones or tablets of fair capacity and memory for its operation.
2. The device must have a camera that allows good-quality images of soil samples for analyses.
3. GPS capability enables the exact and accurate location tagging of soil samples and farmlands, which ensures precise analyses and recommendation generations.

4.2.2 Software Requirements

For the proper operation of AgriLink, these software components will be required:

1. Operating System: AgriLink works on Android (5.0 and above) and iOS (11.0 and above) operating systems.
2. Mobile Application: Users will download and install the AgriLink mobile application from the apk that will be shared by the team.
3. Development Frameworks: AgriLink uses React Native framework that allows cross-platform mobile application development to ensure the appearance and functionality on almost all the mobile devices.
4. Database Management System: AgriLink employs a database management system (DBMS) named MongoDB to store and manage soil data, crop data, user accounts, and other data.
5. Backend Server: The backend server is coded using Node.js which will take care of user authentication, processing of data, and communication with the external APIs.
6. API Integration: AgriLink has integration with external APIs for weather forecasts, market prices, and other open sources of agricultural information. FastAPI is utilized for API development that interfaces with machine learning models running on the server.
7. Machine Learning Libraries: Python-based machine learning libraries such as TensorFlow, scikit-learn, and Keras will be utilized during development and deployment of predictive models for soil classification, crop suggestion, revenue estimation, and loan valuation functionalities.
8. Bot Integration: The Chatbot facility of AgriLink will be integrated with the ChatGPT API for natural language processing and user response handling to user questions on farming practices, pest control, fertilization, etc.

4.2.3 Proposed System Architecture

Figure 4.1 shows the architecture diagram of AgriLink, which can be described as the structural framework of the system whereby different components interact with each other and interact with the modules. It is a representation of how the data pass through the system by defining the relationship of the frontend, backend, databases, and external APIs.

AgriLink combines itself as a network of several interlinked components working together to answer the needs of farmers. At the bottom of it, there is a Node.js backend that acts as the backbone

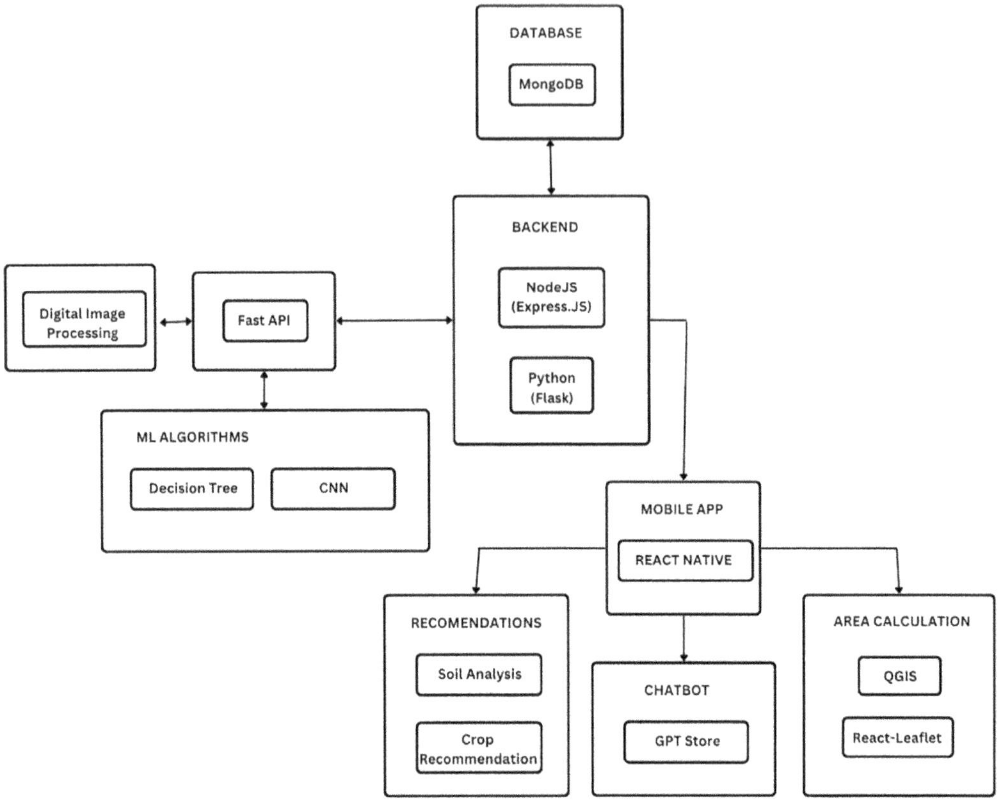

FIGURE 4.1 AgriLink system architecture diagram.

of retrieving and processing data and communicating with various modules. The backend is heavily coupled with a MongoDB database for storing relevant data about soil characteristics, crop details, and user profiles.

From the user's front end, AgriLink is developed based on React Native framework for mobile apps, providing a friendly interface that can be utilized in mobile phones. The mobile app will serve as the main interface for farmers and will thus enable them to take pictures of the soil, view crop suggestions, and use marketplace functions.

The integrated advanced feature of FastAPI contains a link with the Machine learning models deployed for soil categorization and crop suggestion through AgriLink. These models scan the soil images and climate data via intelligence algorithms such as Convolutional Neural Networks and Decision Trees and include a feature for actionable feedback to the farmer.

Moreover, the AgriLink also integrates with the Artificial intelligent chatbot ChatGPT API, which enables farmers to ask questions for farming advice and the language in which they speak. This feature adds value to the perspective of engagement of customers and the response to an instant support platform for crop management, pest control, and fertilizer application across multiple locations queries.

To conclude, the AgriLink architecture is designed for easy integration of various technologies such as machine learning and image processing and chatbots powered by AI for evidence-based insight to farmer for efficient practicing of technology.

4.2.4 Data Flow Model

Dataflow diagrams for AgriLink represent the flow of information through the system, showing how data are passed, manipulated, and transmitted among several modules and components. These diagrams give the viewer a wide overview of how the system works, showing how information flows from input to output.

The Level 0 flow diagram as shown in Figure 4.2 gives general information about the data flows involved in AgriLink, where the major system components, along with their interactions, are presented. It indicates the primary data sources, processing steps, and output destinations with a global view of the movement of information inside the system.

The Level 1 flow diagram as shown in Figure 4.3 examines the dataflow for AgriLink in greater detail. It takes whole parts indicated in the Level 0 diagram and breaks them until even smaller functional subprocesses are distinguished. A data processing step is viewed in greater detail, explaining how data are manipulated and transformed throughout many steps of the operation.

The Level 2 flow diagram, as shown in Figure 4.4, goes into fine detail about the dataflow for AgriLink, focusing upon specific subprocesses shown in the Level 1 diagram. It gives a fine-grained account of the operations that take place as part of any subprocess, retaining traces of data transformations, where modifications and calculations were performed, and interfaces between components.

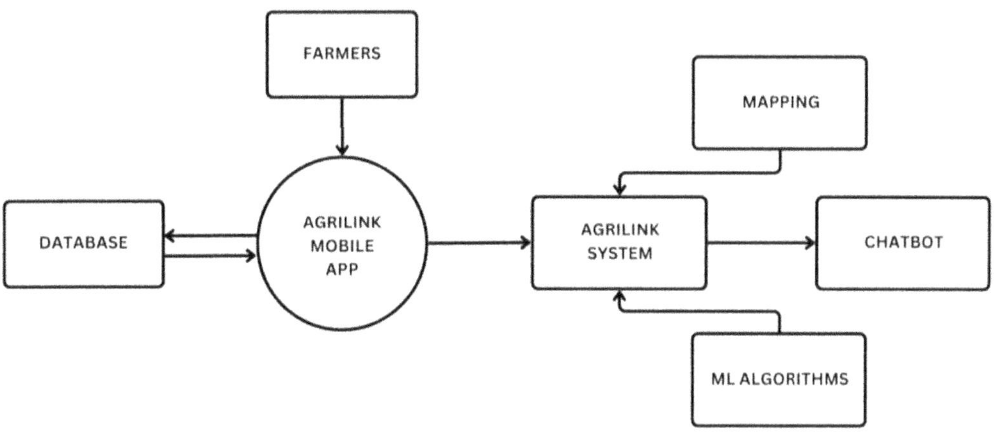

FIGURE 4.2 AgriLink data flow level 0 diagram.

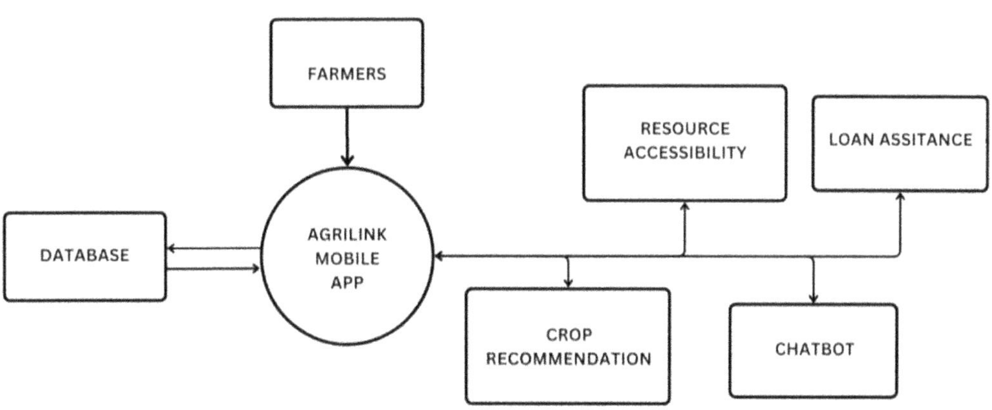

FIGURE 4.3 AgriLink data flow level 1 diagram.

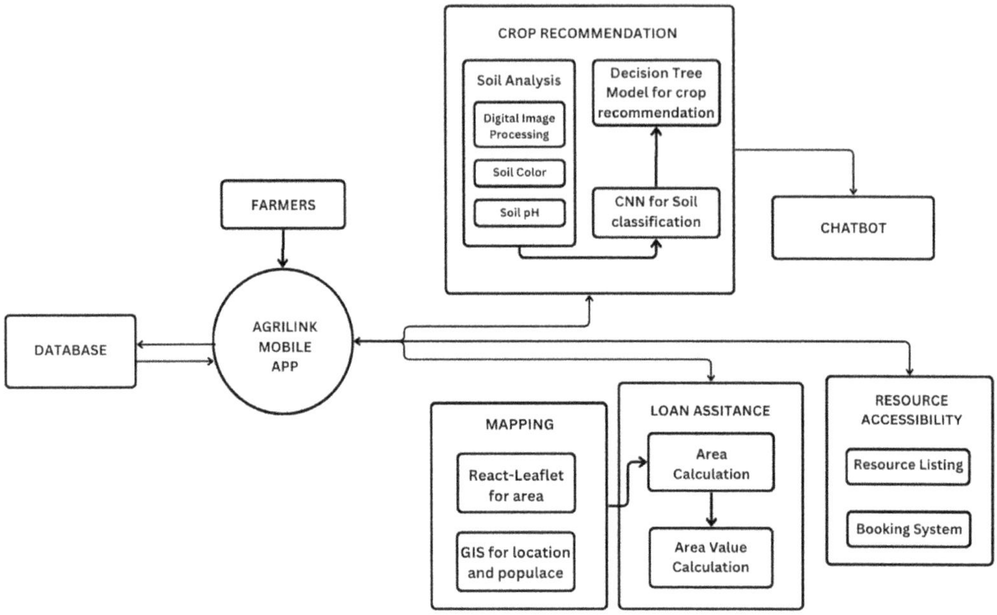

FIGURE 4.4 AgriLink data flow level 2 diagram.

4.2.5 User Interaction Diagram

The user interface diagrams of AgriLink, as seen in Figure 4.5, outline the various interactions existing between the system components and users, indicating how farmers and other agro-stakeholders would interact with the system in performing specific tasks. Such diagrams provide abstractions for system functionalities and user interactions, providing useful information for understanding the requirements for the system and behaviors.

The diagram depicts how the different actors – such as farmers, agricultural experts, and administrators – would interact with different system functions. The blocks of actions would range from soil testing to crop recommendations, loan assistance, and many other tasks.

The soil analysis block is the process where the farmer takes an image of the soil and identifies it with the help of AgriLink mobile application. The crop recommendation use case is a generalized procedure whereby a farmer is advised on what crops to grow depending on the soil analysis and other factors.

Additionally, the user interaction diagram presents the actors' relationships with the system. Specifically, it shows which users have the ability to perform actions on the knowledge domain being deliberated. This allows an overall picture of how AgriLink generally works and how each user must interact with it to achieve their desires.

The user interaction diagram would also be beneficial to several stakeholders because it provides additional specifications on the extent of AgriLink and what it can and cannot accomplish. Hence, adequate communication for the correct requirements analysis in the development of AgriLink.

4.3 IMPLEMENTATION

This section describes how AgriLink is practically brought into existence. It is designed to give the functional processes in implementing the system, starting from an overview of system implementation. After thus, a detailed explanation of the various modules of AgriLink follows, garnering an insight on the algorithms used. Such relationships lead to an understanding of certain code snippets

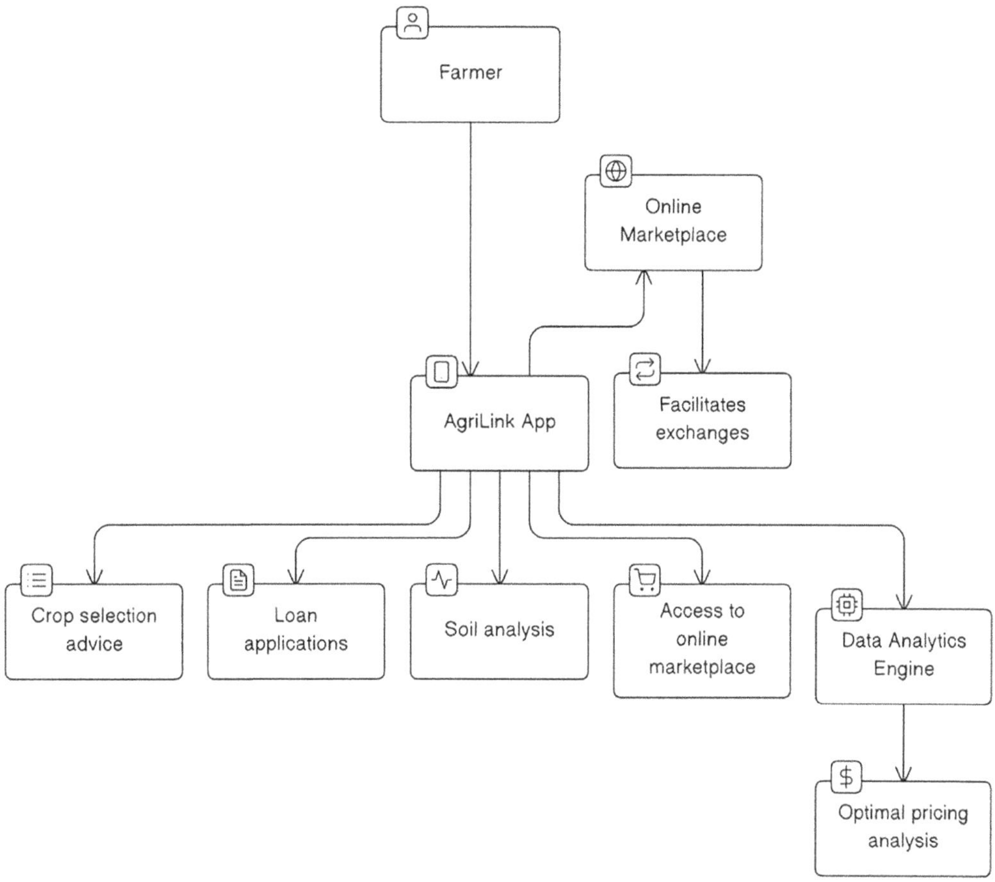

FIGURE 4.5 AgriLink user interaction diagram.

that are the functioning skeleton of the application. By going through this, the user will come to understand how AgriLink operates in the real world.

4.3.1 Overview of System Implementation

AgriLink involves a standard system architecture capable of aggregating various technologies and modules to provide easy-to-use, efficient solutions for farmers.

The mobile application is primarily built on a React Native framework, which enables cross-platform support. The backend was developed using the Node framework. js, which can also create a huge and extendable network to process user requests and track his data. Most used database is MongoDb which makes the system flexible and scalable to handle various types of data.

Key modules of the system are soil classification, crop prediction, estimation of revenue, loan analysis test, market place link and chatbot feature. Each cow welfare module tackles one of the problems Indian farmers face in raising cows.

The soil classification is based on digital image processing and a CNN block that gets the input soil image for classifying it to its suitable soil type. That tells farmers something about their soil and helps them make decisions for how to treat crops.

Decision trees are used in the crop recommendation module which suggests undamaged crops based on their pH value and soil nutrient levels and may be some environmental factors. AgriLink uses artificial intelligence algorithms and personalizes recommendations for agriculture based on location conditions that essentially translates into better profits and yield [3, 10].

The revenue estimation module also depends on various advanced machine learning algorithms for the estimation of crop yield and revenues based on previous crops history and market trend. This will provide the farmers suitable financial planning and risk management for all parameters of crop cultivation [6].

Also, AgriLink creates a market place feature that allows farmers to buy, sell, and rent agritools or products. Thus, it promotes collaboration and sharing of resources in the pure agriculture community, where farmers serve each other with their combined knowledge. Chatbot functionality gives farmers help all the time about all questions, regarding farming techniques, pest control, and quite more [5].

In summary, while the system implementation of AgriLink happens to portray a more approachable yet comprehensive platform setup designed to equip farmers with data-driven knots between solutions that increase productivity and profitability while ensuring sustainable advancement in the field of agriculture.

4.3.2 Modules Description

AgriLink consists of several modules, each dedicated to a particular issue that farmers face to find a complete solution to the problem. Let me break down each module:

1. Soil Classification Module:
    - This module works based on digital image-processing techniques and a convolutional neural network (CNN) model to analyze soil images captured by farmers.
    - In this module, the CNN model employs color-oriented characteristics extracted from images to classify soil types.
    - It basically provides useful information about the soil to farmers, which helps in crop selection and management decisions.
2. Crop Recommendation Module:
    - The crop recommendation module aims to recommend suitable crops, Decision Tree algorithms in particular, depending on soil nutrient levels, pH values, and environmental factors.
    - It will consider temperature, rainfall, and humidity when providing customized crop recommendations according to individual farming conditions.
    - This module assists farmers by optimizing their crop selection for maximum yield plus overall farm productivity.
3. Area Cost Estimation Module:
    - Geographic Information System (GIS) technology enables this module to accurately calculate the area of the farmland for farmers.
    - Farmers mark the boundary of their land on the digital map interface of the module, and the system then calculates the area based on the marked coordinates.
    - The module fetches government-set rates per unit area (circle rates, etc.) and based on the calculated area can estimate the value of the farmland.
    - It helps in valuing land property and providing valuable insight for financial planning and loan appraisal.

- Integrating the area calculation and estimation of costs enhances this module's overall toolkit for financial management, empowering farmers with better decision-making in crop farming activities.

4. Marketplace Integration Module:

   - AgriLink combines a marketplace feature where farmers can buy, sell, or rent agricultural tools, products, and produce.
   - Helps create collaboration and space for resource sharing among the farming community so that the farming resources can be accessed efficiently while providing economic opportunities to farmers.
   - The marketplace module boosts community participation and sustainable farming practices.

5. Chatbot Integration Module:

   - The module runs a natural-language-processing chatbot for providing real-time support to farmers.
   - It answers questions ranging from farming practices, pest control, market trends, and other horticultural problems.
   - It enhances user involvement and aids the farmers in getting the information and support they need for improved farming practices.

Each module is instrumental in continuously supporting the farmers with progressive insights and solutions toward data for productivity, profitability, and sustainability in agriculture.

4.3.3 Algorithms for the Methodology

AgriLink uses various algorithms that aid in agricultural data analysis to provide actionable insights to farmers. Digital image processing, convolutional neural networks (CNN), and Decision Tree algorithm hold great relevance among these algorithms.

Digital image processing algorithms are employed to capture relevant information from soil images taken by digital cameras or imaging devices. In image processing, the considerable enhancement of picture quality, noise removal, and the acknowledged viewpoints such as texture and soil color help evaluate the image for further analysis.

- Digital Image Processing for soil-type classification:

  1. To enhance image quality by eliminating noise and normalizing the lighting conditions of the photographic image.
  2. Improving image quality by contrast enhancement, histogram equalization, and filtering.
  3. Image conversion to grayscale or other color space if required.
  4. Soil properties identification of value for future analysis, which includes texture, color, and structure.
  5. Detection of significant features using edge detection, thresholding, and morphological processing.
  6. Segmentation techniques are utilized to isolate the soil areas from background and other objects.
  7. Train the CNN model using a dataset of images of soils labeled with their corresponding soil types.

8. Apply the trained model by classifying new soil type images based on feature extraction.9. Verify the classification output based on accuracy measures and through visual inspection.
10. Generate classified images of soils with identified or marked soil type(s).
11. Feed the soil classification output to the soil classification and crop recommendation modules of AgriLink.

CNN algorithms are used for soil classification based on processed images. Models trained on labeled soil image datasets can classify soil types effectively using precise identification of soil properties and characteristics that are significant for crop cultivation [1, 2].

- Convolutional neural network (CNN) for soil classification:

1. Preprocess soil images to enhance their quality and remove noise.
2. Extract features from the preprocessed images using convolutional layers.
3. Apply ReLU activation function to introduce non-linearity.
4. Downsample feature map using pooling layers to retain useful information.
5. Regularize the network using dropout layers in order to mitigate overfitting.
6. Flatten the output from convolutional layers into a one-dimensional vector.
7. Classify soil types on the basis of features extracted using fully connected layers.
8. Train the CNN model through forward propagation and backpropagation using a minimization of errors produced.
9. Provide soil classification result to soil classification module and crop recommendation module in AgriLink.

Decision Tree algorithms provide recommendations on crops that can be grown based on nutrient levels, weather observations, and geographical considerations. They will recursively partition the dataset to arrive at a conclusion regarding which crops will grow best in those conditions, hence providing farmers personalized, data-driven knowledge to nurture and optimize their agricultural practices [1, 4].

- Decision Tree in Crop Recommendation:

1. Prepare up a dataset, which consists of soil nutrient levels, weather conditions, and crop attributes.
2. Using feature values nitrogen level, phosphorus level, pH value, and temperature, split the dataset into subsets.
3. Determine the best feature to split on using criterion i.e., information gain.
4. Recursively split the subsets until the leaf nodes indicate individual crops.
5. Assign the leaf nodes with the best crop according to a majority voting system.
6. Provide soil classification results to the crop recommendation module in AgriLink.

The fusion of technologies in AgriLink empowers farmers with modern technological solutions for soil analysis, crop recommendation, and yield prediction, which can help with productivity and sustainability in agriculture.

4.4 RESULTS

4.4.1 Code Snippets

In this section, we will present several snippets from the extensive software development of the AgriLink application and API systems. From CNN layers – Figure 4.6 – to several languages support – Figure 4.7 – and to the interface for logging in – Figure 4.8, the source code snippets

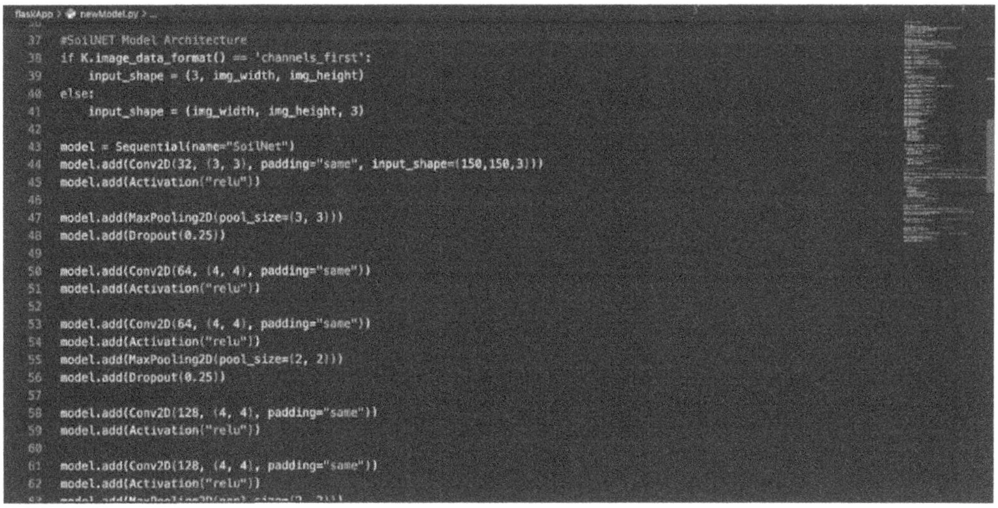

FIGURE 4.6 Code snippet with CNN deep learning model layers.

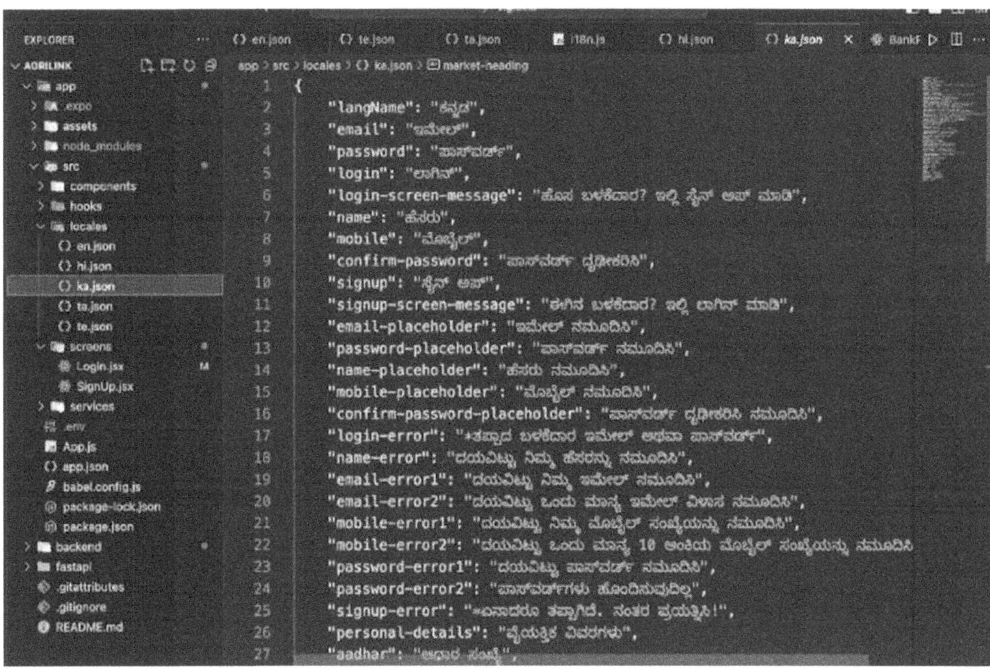

FIGURE 4.7 Code snippet from multiple language support code.

give insight into the implementation of AgriLink features that span elaborate machine learning algorithms and user interface interactions.

4.4.2 Testing

Testing is quite fundamental to any software development process. Importantly, it is the PARAMOUNT way of ensuring correctness, functionality, and performance aspects of that APPLICATION. For AgriLink, testing ensures the module, the algorithms, and even user interactions are assured and

FIGURE 4.8 Code snippet from login user interface code.

confirm that the application works fine, Just as the user may want it to w. The following is a summary of the test procedures conducted in AgriLink in ensuring QUALITY, CHECK, AND ROBUSTNESS are in place: (show how did u exercise the test whether it will be unit level or integration).

4.4.2.1 Unit Test Cases

Unit testing may be stated as the verification of individual components or units of code to make sure they are working correctly and appropriately. Nonetheless, unit testing is a testing first or foremost method which is applied to verify certain modules, algorithms, and functions of the application. The following subsections illustrates Unit testing protocols used In AgriLink, soil classification module testing, Crop recommendation algorithm, land area calculation module, land area estimation module, market integration, and chatbot interactions.

- Sample Unit Test Cases for the Soil Classification Module:

  AgriLink's Soil Classification module is a core application component that is directly responsible for identifying soil types from input images with precision. The test cases below ensure that the functionality of the module exhibits stability and reliability. The tests explore a variety of possibilities, including soil classification and invalid inputs, as presented in Table 4.1.

- The Crop Recommendation Module unit test cases:

  This provides a glimpse of how the module is used to support farmers with data-driven insights about suitable crops based on soil characteristics. The unit test cases like the ones below validate this module's correctness and reliability. These tests evaluate the decision tree prediction, input validation, and treatment of outliers to make sure precision in recommending crops according to Table 4.2.

TABLE 4.1
Soil Classification Module Unit Test Cases

TC No.	Test Case Name	Input	Expected Output
1	Soil classification	Pre-processed clayey soil image	Predicted as clayey soil
2	Soil classification	Pre-processed alluvial soil image	Predicted as alluvial soil
3	Invalid input	Non-image input	Error: image required

TABLE 4.2
Crop Recommendation Module Unit Test Cases

TC No.	Test Case Name	Input	Expected Output
1	Crop recommendation	N: 25, P:50, K:25, pH:6, temp:24, humidity:60, rainfall:100	Recommended crop: peas
2	Input validation	Missing input data	Error: required field
3	Outlier handling	Extreme values in input	Do not allow input

- Unit Test Cases for Land Area Calculation Module:

  Correct estimation of land is the main function of many aspects of agricultural planning. The land area calculation module in AgriLink uses GPS coordinates and geometry algorithms to accomplish this task. The test cases presented here would enable unit testing for correctness of area calculation, marker placement functionality, and error handling, as shown in Table 4.3.

- Marketplace Integration Module Unit Test Cases:

  The Marketplace Integration module facilitates resource sharing and trading by enhancing collaboration among farmers. The unit test cases below validate some core features of this module which include creation of listings, searching capability, and transaction handling. They provide an assurance of seamless user experience and consistent operation of the marketplace, as in Table 4.4.

AgriLink intends to perform complete unit tests to early locate and fix any bugs or issues within the application, thus assuring the quality and robustness of the application.

4.4.2.2 Integration Test Cases

Integration testing is an integral and critical phase within the software development life cycle. Integration testing targets the interrelationships and integration networks between the modules, components, and systems within the application. In AgriLink, integration testing tests communication and interoperability between different modules, including soil classification, crop recommendation, land area calculation, land value estimation, marketplace integration, and chatbot functionality. This section provides an overview of the integration testing process conducted on AgriLink, including testing end-to-end workflows, data transfers among modules, integration with external services (i.e., weather APIs and market rate APIs), user interaction, interface integration, scalability, performance, error handling, security, and data privacy.

Table 4.5 indicates further tests that were performed on the ArgiLink app to verify its integration and that it provided a seamless experience for the user.

TABLE 4.3
Land Area Calculation Module Unit Test Cases

TC No.	Test Case Name	Input	Expected Output
1	Marker placement	Mark area on map	Four markers on the map interface
2	Area calculation	GPS coordinates: [{"latitude": 12.97364, "longitude":77.59300}, {"latitude": 12.97370, "longitude":77.59332}, {"latitude": 12.97331, "longitude":77.59299},{"latitude":12.97331, "longitude": 77.59332}	Calculated land area: 1912.083 sq. ft.
3	Error handling	Incomplete marker placement	Error: Requires 4 Markers

TABLE 4.4
Marketplace Module Unit Test Cases

TC No.	Test Case Name	Input	Expected Output
1	Listing Creation	Product type Fertilizers	Product listed under Fertilizers
2	Contact Details	Enter contact number for the product	Contact number visible under the correct product

TABLE 4.5
Integration Test Cases for AgriLink

TC No.	Test Case Name	Input	Expected Output
1	End-to-end workflow integration	Simulated soil image data, weather data, user location	Successful exchange of data between modules, accurate crop recommendation based on analysis
2	Data-flow integration	Sample soil nutrient data, pH levels, weather conditions	Proper integration between modules, accurate crop suggestions
3	Integration with external services	Request to fetch real-time weather data and market rates	Successful retrieval of data from external APIs, accurate incorporation into AgriLink system
4	User interaction and interface integration	User requests for crop recommendation and marketplace access	Smooth communication between frontend UI and backend services, correct responses to user queries
5	Scalability and performance testing	Simulated concurrent user requests and data processing	System maintains responsiveness and stability under heavy loads, minimal performance degradation
6	Error handling and resilience testing	Intentional disruption of backend services or invalid inputs	Graceful handling of errors, meaningful error messages, system recovery without data loss
7	Security and data privacy testing	Attempted unauthorized access to sensitive user data	Prevention of unauthorized access, encryption of sensitive data, adherence to data privacy regulations

AgriLink is an application through which seamless and reliable user interference is afforded to farmers undergraced by an excellent system of guidance and recommendations for best agricultural practices.

4.4.3 User End Results

The results demonstrating the implementation of the AgriLink are provided in this section. The results indicate the success of the system in soil classification, crop recommendation, calculation of land area, and marketplace integration. These findings enhance AgriLink's credibility for practical and reliable decision-making for the farmers.

AgriLink achieved remarkable results in several fields of agri-management, as it has redefined traditional agricultural practices. Its aerial strength is in soil classification, based on the deployment of digital image processing with advanced convolutional neural network (CNN), and also these features have, in fact, classified soil types on processable characteristics derived from soil-based images as depicted in Figure 4.9. The analysis from AgriLink allows farmers to know their soil better and insightful decision making about choice selection, fertilization, and land management.

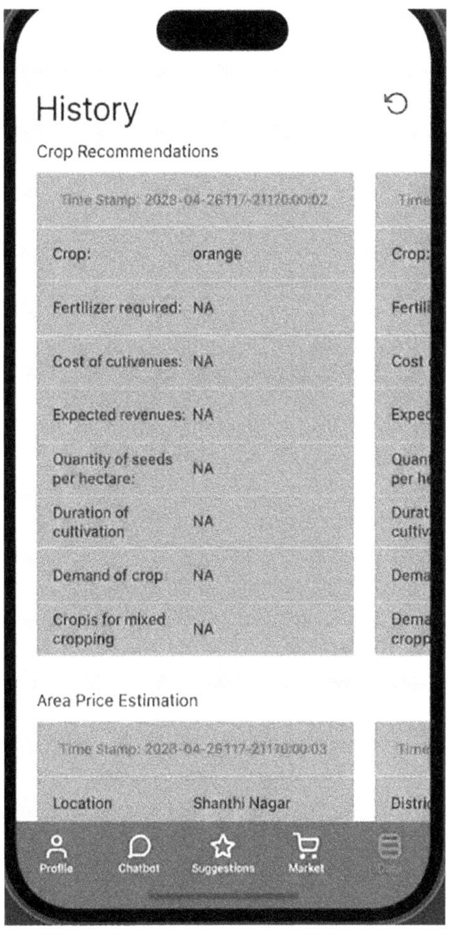

FIGURE 4.9 Crop recommendation results.

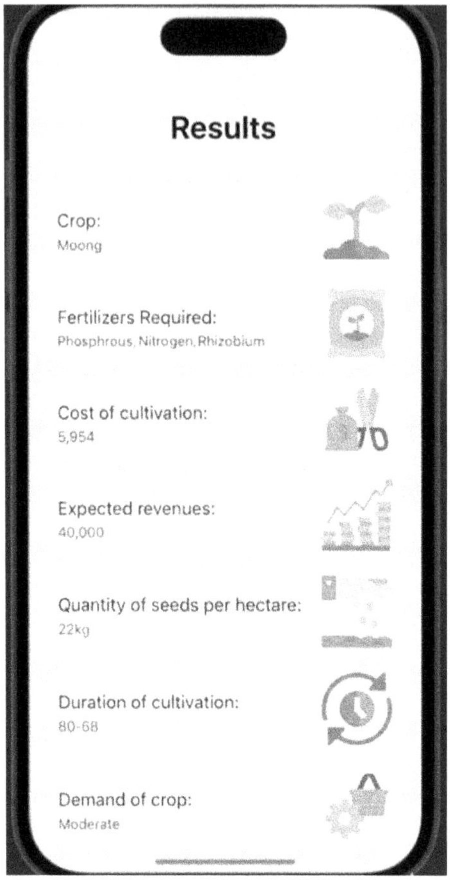

FIGURE 4.10 History of recommended crops.

Most notably, the crop recommendation module of AgriLink, as seen in Figure 4.10, was effective in aiding farmers to make crop selections. The analysis of soil nutrients, pH, and climate conditions by AgriLink provides crop recommendations personalized for the best yield potential for farmers. The system would ensure that farmers choose crops with optimized chances of good harvest and good profits thus optimizing their yield and returns.

Aside from crop management, AgriLink does a very good job in financial analysis. Working out the square meter area of farmland and the approximate fair market value for it, as shown in Figure 4.11, AgriLink enables farmers to make informed financial decisions and efficiently invest in their farming looking to judge their financial health and suitability for loans as shown in Figure 4.12.

Another way AgriLink steps up its value proposition is the community marketplace feature demonstrated in Figure 4.13. It enables farmers to share resources and trade equipment or even find out pricing for a particular product available in the market. Such platform promotes knowledge exchange among the agriculture community, eventually increasing production and making the agricultural practice sustainable.

After all, the chatbot interface is quite user-friendly and assists farmers utilizing this interface in seeking immediate help and guidance for various farming-related questions. The AI-driven chatbot interface enlightening users and spreads knowledge will greatly benefit Indian farmers.

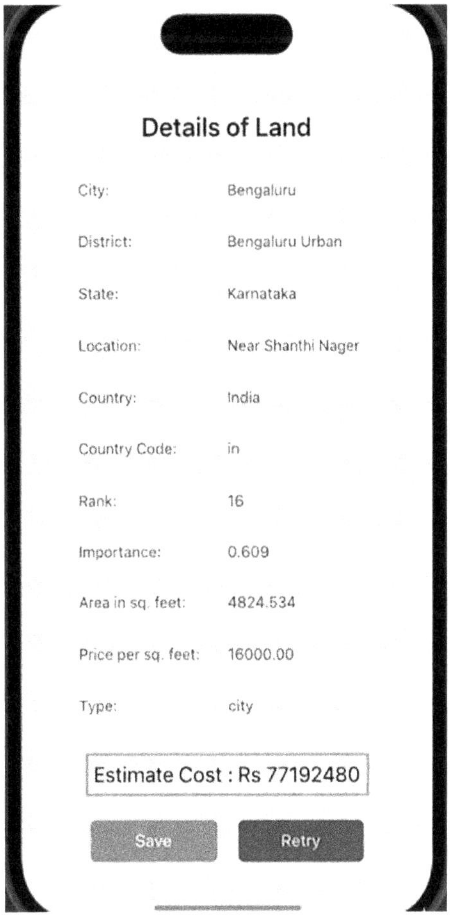

FIGURE 4.11 Land cost estimation and related details.

AgriLink's outputs, in summary, show how the platform has revolutionized agriculture through provision of farmers with tools, insights, and resources to boost productivity, profitability, and sustainability of agricultural operations.

4.5 CONCLUSION

The section concludes with a summary of some of the critical insights and results of AgriLink regarding how it could revolutionize Indian agriculture. In conclusion, AgriLink is a one-stop solution that will revolutionize farming all over India. It is a melange of digital image processing, machine learning, and artificial intelligence that work around the clock to solve critical farmers' challenges at every stage from the soil to the finance issue. Assumptions on the soil classification and recommendation of the plant to crops were made, helping the farmer make learned decisions on the type of crops to grow, thus increasing productivity and profitability. Additionally, AgriLink helps in the financial aspect through profit forecasting and loan eligibility analysis, hence increasing the farmer's knowledge and ease to credit and making farming more sustainable. The community marketplace creates a platform for farmers to collaborate hence makes farming activities more

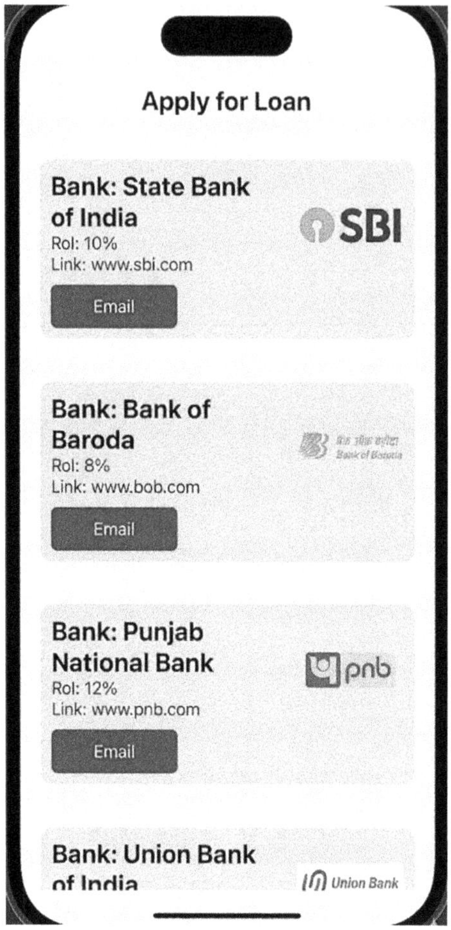

FIGURE 4.12 List of banks to email financial analysis.

efficient and resilient. Through an intuitive interface, and a friendly chatbot to engage the farmers, AgriLink will empower any farmer regardless of the level to spread useful knowledge and practice. In the ever challenging world, AgriLink represents progress, moving from agri-etopia, metaphorically illuminating the road ahead through a changing landscape to the future, making the implementation of tem technologies less cumbersome. The trailblazing's journey in AgriLink's development and scalability offers a lot of hope for farming in India's future. If AgriLink can grow, community, and find a place in the entire agricultural value chain, it can transform and empower millions of farmers to thrive in an increasingly challenging and ever-changing landscape. In brief, AgriLink represents an age of transformation for the mindset of agricultural management – an ownership embracing precision-farming, finance empowerment, and bonding society. Equipped with the cap abilities to realize agricultural sustainability and food security in India, AgriLink is eager to usher a better India tomorrow for our farmers.

FIGURE 4.13 Marketplace listing items and owners contact information.

REFERENCES

[1] K. Varsha, G. Prakasha, N. Sridevi and J. Venkatagiri, "Soil Classification, Crop Selection and Prediction of Fertilizer Based on Soil Series," *2022 International Interdisciplinary Humanitarian Conference for Sustainability (IIHC)*, Bengaluru, India, 2022, pp. 1174–1178, doi: 10.1109/IIHC55949.2022.10059636.

[2] S. Bhanumathi, M. Vineeth and N. Rohit, "Crop Yield Prediction and Efficient Use of Fertilizers," *2019 International Conference on Communication and Signal Processing (ICCSP), Chennai, India*, 2019, pp. 0769–0773, doi: 10.1109/ICCSP.2019.8698087.

[3] P. S. Nishant, P. Sai Venkat, B. L. Avinash and B. Jabber, "Crop Yield Prediction Based on Indian Agriculture Using Machine Learning," *2020 International Conference for Emerging Technology (INCET)*, Belgaum, India, 2020, pp. 1–4, doi: 10.1109/INCET49848.2020.9154036.

[4] M. Kalimuthu, P. Vaishnavi and M. Kishore, "Crop Prediction Using Machine Learning," *2020 Third International Conference on Smart Systems and Inventive Technology (ICSSIT)*, Tirunelveli, India, 2020, pp. 926–932, doi: 10.1109/ICSSIT48917.2020.9214190.

[5] R. Medar, V. S. Rajpurohit and S. Shweta, "Crop Yield Prediction Using Machine Learning Techniques," *2019 IEEE 5th International Conference for Convergence in Technology (I2CT)*, Bombay, India, 2019, pp. 1–5, doi: 10.1109/I2CT45611.2019.9033611.

[6] G. Thapaswini and M. Gunasekaran, "A Methodology for Crop Price Prediction Using Machine Learning," *2022 IEEE 2nd International Conference on Mobile Networks and Wireless Communications (ICMNWC)*, Tumkur, Karnataka, India, 2022, pp. 1–7, doi: 10.1109/ICMNWC56175.2022.10031852.

[7] K. Ashwitha, A. Kumar, B. Rao, P. Salian and A. P. Supravi, "An Approach to Predict Loan Eligibility Using Machine Learning," *2022 International Conference on Artificial Intelligence and Data Engineering (AIDE)*, Karkala, India, 2022, pp. 23–28, doi: 10.1109/AIDE57180.2022.10059881.

[8] S. M. Pande, P. K. Ramesh, A. Anmol, B. R. Aishwarya, K. Rohilla and K. Shaurya, "Crop Recommender System Using Machine Learning Approach," *2021 5th International Conference on Computing Methodologies and Communication (ICCMC)*, Erode, India, 2021, pp. 1066–1071, doi: 10.1109/ICCMC51019.2021.9418351.

[9] G. Mariammal, A. Suruliandi, S. P. Raja and E. Poongothai, "Prediction of Land Suitability for Crop Cultivation Based on Soil and Environmental Characteristics Using Modified Recursive Feature Elimination Technique With Various Classifiers," in *IEEE Transactions on Computational Social Systems*, vol. 8, no. 5, pp. 1132–1142, Oct. 2021, doi: 10.1109/TCSS.2021.3074534.

[10] R. Gupta et al., "WB-CPI: Weather Based Crop Prediction in India Using Big Data Analytics," in *IEEE Access*, vol. 9, pp. 137869–137885, 2021, doi: 10.1109/ACCESS.2021.3117247.

5 Influence of Internet of Things and Artificial Intelligence in Precision Agriculture of Coffee Crop

Jagadamba G, Chayashree G, Hemavathi, and Varun Jayadeva

5.1 INTRODUCTION

The traditional agriculture methods have moved to precision agriculture, utilizing IoT sensors, drones, and satellite images. However, inclusion of AI employed for real-time monitoring and predictive decision-making made the agricultural farming robust and scalable. Thus, optimizing the precision agriculture is **essential** in a site-specific crop management through systematic data collection and analysis, enabling informed decisions tailored to specific land characteristics. The whole process encompasses the need of proper data collection, data analysis, input planning, resource application, crop mapping, and management. The discussions on these phases are as follows.

Data Collection Phase: In this phase, sensors, drones, GPS (Global Positioning System) units, and other IoT devices collect data. The data collection includes soil moisture, crop health, temperature, etc. from the field. These data become the foundation for every subsequent activity, such as analysis, planning, and execution.

Data Analysis Phase: The phase links the gap between raw sensor data and informed decision-making. Once these data are gathered, it is preprocessed, integrated, and interpreted to uncover patterns and insights that can optimize farming practices followed by Sahu et al. (2026) and Singh et al. (2024).

Inputs Planning Phase: This phase decides the application of resources, such as seeds, water, fertilizers, and pesticides, are optimized using insights derived from IoT data and AI analytics. Many researchers including Ravi et al. (2022), and Kumar et al. (2023) well-defined the objectives to deliver proper input, at the appropriate time and place. This increased the crop productivity while minimizing the costs and environmental impacts.

Resource Application Phase: This phase focuses on the efficient and targeted use of agricultural inputs. The proper usage of fertilizers, water, pesticides, and energy is based on data-driven insights and AI analysis as mentioned by Mulla (2013), Shamshiri et al. (2018)., Wolfert et al. (2017). This phase helps in reducing waste and environmental impact while guaranteeing that crops receive what they need. More autonomous, sustainable, and economically viable resource application is required for resource efficiency, crop health, and yield quality management.

Crop Mapping Phase: This phase used remote sensing technologies and AI algorithms defined by Zhang et al. (2022), Xie et al. (2008), and Kamilaris & Prenafeta-Boldú (2018). These works used geospatial data to create detailed maps of crop types, health, and growth stages across agricultural fields. Thus, crop mapping allows farmers to monitor crop development

DOI: 10.1201/9781003613510-5

TABLE 5.1
Comparison of Traditional and Precision Agriculture Practices

Feature	Traditional agriculture	Precision agriculture
Data collection	Manual	Sensor-based
Decision making	Experience-based	Data-driven
Resource mapping and utilization	Uniform	Site and application-specific
Environmental impact	Higher	Lower

accurately. It is essential for pre-forecasting, such as yield forecasting, pest detection, and planning precise interventions.

Evaluation Phase: This phase involves analyzing the effectiveness of implemented practices and technologies. Evaluation ensures continuous improvement in precision agriculture by validating technological applications and guiding future innovations.

A comparison of traditional and precision agricultural practices has been summed up in Table 5.1 to give a precise introduction.

The chapter is organized into five sections. Section 5.1 introduces the concept of precision agriculture with IoT and AI. Section 5.2 introduces the IoT and AI role in precision agriculture. Section 5.3 defines the integration of IoT and AI. Section 5.4 introduces the case study and its impact on precision agriculture, and Section 5.5 concludes the chapter with future possibilities to improve precision agriculture.

5.1.1 Role of IoT and AI in Precision Agriculture

The role of IoT in precision agriculture has existed for a long time and has been improved in many ways for disease monitoring. Sensors track field conditions favorable to diseases (e.g., high humidity, low temperature). Drones and cameras capture leaf imagery for AI analysis. A study by Malik, Arvind, and Ramachandra Kumar (2021) presents the ability of IoT in agriculture applications. The usage of sensors for soil preparation, monitoring crop conditions, irrigation management, and detecting insects and pests is always a great resource for precision agriculture.

AI processes the massive data sets generated by IoT devices using machine learning models, computer vision, and many decision support systems for precision agriculture. However, machine learning models are especially used for predicting the crop yield, disease outbreaks, and pest infestations. However, computer vision methods are used to analyze images from drones for plant health and growth stage assessment. Many decision support systems provided actionable recommendations for planting, fertilizing, and harvesting.

5.1.2 Integration of IoT and AI

IoT plays a foundational role in enabling AI-based disease prediction and monitoring by timely and accurate assessments. However, data collection happens through temperature, humidity, soil moisture, and leaf wetness sensors. Imaging devices such as drones and stationary cameras capture real-time images of crop foliage. These data streams are essential inputs for AI models for training and detection of early warning signs.

IoT-enabled systems transmit data to cloud platforms or edge processors and are the sources for monitoring, detecting, and alert systems. For example, if certain conditions match known disease-favorable patterns (e.g., high humidity + low temperature), alerts are automatically triggered, giving farmers time to act before symptoms appear.

Finally, we can conclude that IoT devices collect data, and AI analyzes and predicts risk in precision agriculture. Thus, the synergistic approach is that by combining IoT and AI creates an intelligent ecosystem for predictive analytics (e.g. forecasts on real-time data), automated decision-making (e.g. minimizes human intervention), and Feedback Loops (e.g. Systems learn and improve over time).

5.1.3 Case Study and Applications

Thus, AI analyzes temporal and image data to forecast disease outbreaks and suggest preventive actions. Deep learning, particularly CNNs, excels at recognizing visual disease patterns in crops. In this section, we considered the case study of coffee leaf disease detection using CNN implementation and its outcomes using IoT for precision agriculture. A work is executed using the CNN method by Anudeep et al. (2024) which detected early disease detection and intervention, reduced crop loss, and improved precision in agrochemical application, with improved sustainability.

Figure 5.1 shows the block diagram of the proposed coffee leaf disease detection system that includes five modules: Dataset of coffee leaf, data pre-processing, data labelling, feature extraction, and disease recognition.

The coffee leaf dataset consisted of images of coffee leaves that were sourced from ScienceDirect, Mendeley, and images extracted through a Samsung S1 mobile camera. ScienceDirect contains images of leaves infected with leaf miner, cercospora, and rust, while Mendeley data consists of images of both healthy and unhealthy coffee leaves. The images collected from Samsung mobile camera included both healthy, cercospora, and rust images of Arabica and Robusta Coffee leaf.

The collected images are preprocessed to ensure consistency in size and format as shown in Figure 5.2. After data preprocessing, each image in the dataset is labeled and classified as Healthy or diseased leaf. Under diseased leaf, we are classifying it as either to Leaf miner, Cercospora, or Rust. Proceeding to labeling, feature selection is done using a CNN to examine coffee leaf images affected by diseases.

The proposed system is tested on datasets from Mendeley data (n.d.) and on an own collection of images of Arabica and Robusta Coffee leaves. It consisted of 2000 images and 1800 images, of which are related to rust and healthy, and another dataset includes images of miner and cercospora diseases. These two datasets were combined as one so that we can efficiently manage the different diseases in a single unit. Disease recognition is done through phases, starting with model training by dividing the available dataset into a training subset, a validation subset, and a testing subset. The

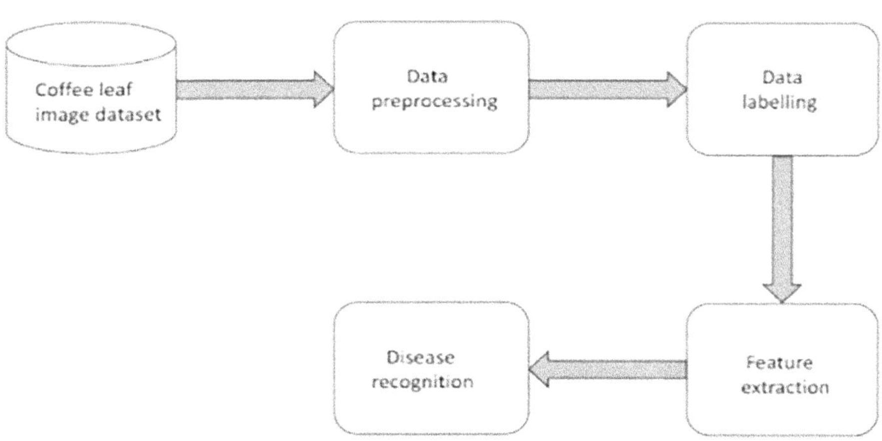

FIGURE 5.1 Coffee Leaf Disease Detection System.

FIGURE 5.2 Leaf Diseases of Leaf miner, Cercospora, and Rust.

typical split done here is 80% for training and 10% for testing, and another 10% for validation. However, the overall testing can be mentioned as 20%. This partitioning allows model training on a majority of the data, validation to tune hyperparameters and evaluate performance during training, and testing on unseen data.

Further, the total number of samples was divided into two folders, namely training and testing. In the initial attempt, 60% of the dataset was used for model training, and the remaining 40% was used for the model's performance testing. From the ratio of testing, an accuracy of 88% was seen, and however, the model might encounter challenges in generalizing to unseen data. By looking into overfitting and underfitting scenarios, the 70% dataset was allocated for training and 30% for testing, resulting in an improved accuracy of 94.6%. However, the diverse testing set helps to generalize exceptionally well to unseen data. Hence, analysis for improved accuracy can be seen with 80%–20% split. Thus, the model performs at its best on the 80%–20% split due to improved generalization and better representation of the true model performance on new, unseen samples. Adjusting the training and testing set proportions helped in improving the model's accuracy. The shift from 60%–40% to 70%–30% allowed the model to generalize better, and the final 80%–20% split provided the most representative evaluation, resulting in the highest accuracy of 97.37%.

Nearly 3000 images were taken for training and 800 images for the testing process. Further self-developed datasets were provided for the testing phase. The testing of the dataset was personally gathered from a coffee estate in the nearby Hassan district of Karnataka state. This data collection process could involve physically visiting coffee plantations, interacting with the coffee plant owners, and collecting various pieces of information about the coffee leaves. Nearly 250 images were taken for this, and the leaf structure varies for different seasons. After testing and training, the result obtained from the experiment shows an increase in accuracy and a decrease in loss as the number of epochs increases. Totally 30 rounds of epochs were done, and finally, the system achieved an accuracy of 97.37% with 7.19% training loss. However, a test accuracy of 97.92% is achieved with 5.25% test loss.

A few observations were conducted during experiments and noted below for different disease prevalence and the impact of rainy and dry seasons.

Coffee Leaf Rust disease: During the rainy season, coffee leaf rust disease thrives in wet and humid conditions. During this season, humidity is high, and there will be frequent rainfall; the spores of the coffee leaf rust fungus can germinate and spread rapidly. Warm temperatures combined with moisture create an ideal environment for disease development. However, due to reduced humidity and less frequent rainfall, the spread of rust is slow. However, if the disease has already been established during the rainy season, its effects might still be visible during the dry season. The fungus might survive on infected leaves or debris.

Coffee Leaf Miner diseaset: The coffee leaf miner disease is a pest occurring in the rainy season, rather than a disease. However, heavy rain can wash away eggs and larvae, reducing infestations. But increased humidity during the rainy season might create favorable conditions for the pest to reproduce. In drier conditions, the activity of the coffee leaf miner will decrease, resulting in lower infestations. However, if the pest has laid eggs before the dry season, its impact might continue, and we can observe the effect.

Cercospora Leaf Spot disease: Cercospora leaf spot disease arrives in moist conditions, especially during rainy seasons with high humidity. Spores are easily spread by rain splashes and wind and low down during the dry season due to reduced humidity. However, if the disease has established during the wetter months, its effects could still be visible on infected leaves. The behavior of these diseases can also be influenced by local climate, altitude, temperature, and the specific coffee varieties being grown. Additionally, the management practices implemented by coffee growers, such as pruning, sanitation, use of fungicides or pesticides, can also impact the severity and spread of these diseases.

From the experimentation and AI analysis, accuracy, precision, recall, and score parameters for CNN are depicted graphically in Figure 5.3. In the same line, Figure 5.4 provides an insight into the experimental approach of training and validation for different epochs. From the analysis, it was observed that after 7 epochs, the system behaved appropriately or hand-in-hand.

In a similar way, the training and testing loss was also studied and can be seen in Figure 5.5 for different epochs, and the loss stabilized after 5 epochs.

A Flask application was used to open a web browser for real-time experimentation. Flask is necessary for a lightweight web framework, enabling easy deployment of machine learning models as web services. Once the Flask application is running, it opens a web browser with the local address. A web interface for all the analysis is ready and can be seen in Figure 5.6. From all the testing methods, it is clear that CNN is the best classifier for coffee leaf disease detection/prediction.

The application of IoT and Artificial Intelligence (AI) in coffee leaf disease detection has shown promising results in automating the identification and diagnosis process. By leveraging CNN, the proposed system has the potential to increase the efficiency and accuracy to 97.37%, and enable timely interventions to prevent crop losses and improve coffee production.

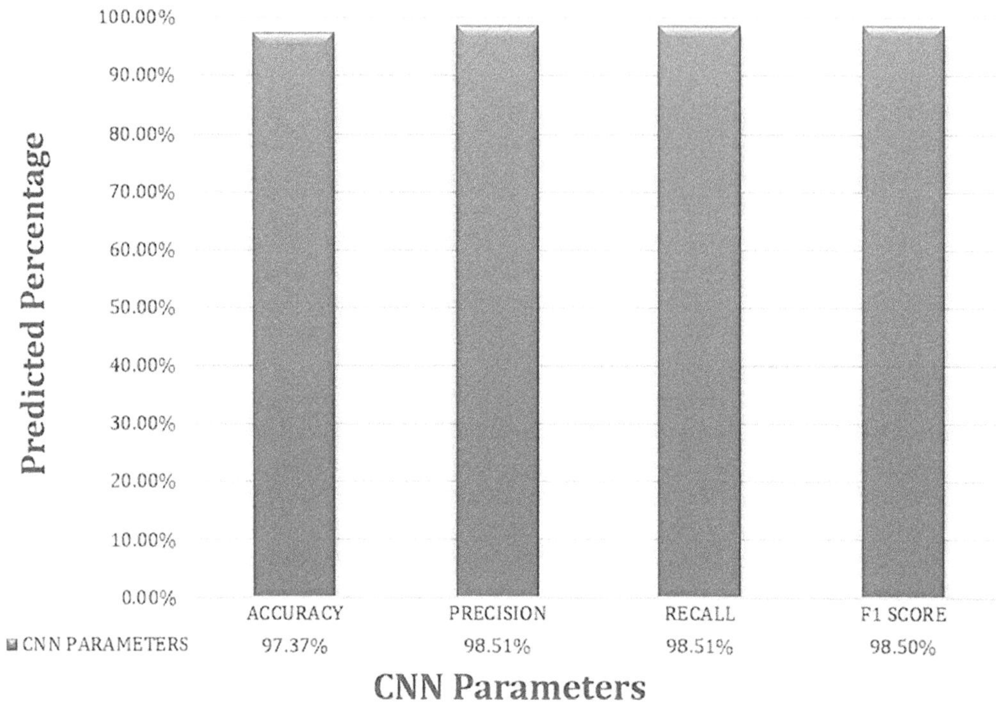

FIGURE 5.3 CNN Parameters.

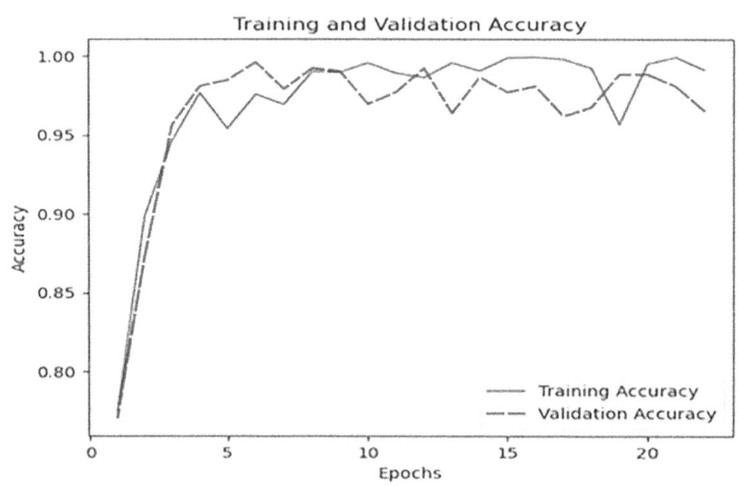

FIGURE 5.4 Training and Validation Accuracy for different Epochs.

Apart from all these benefits, a challenge of high initial investment in setting up data collection through IoT architecture and application development is always expected, and there is a need for technical expertise to use it.

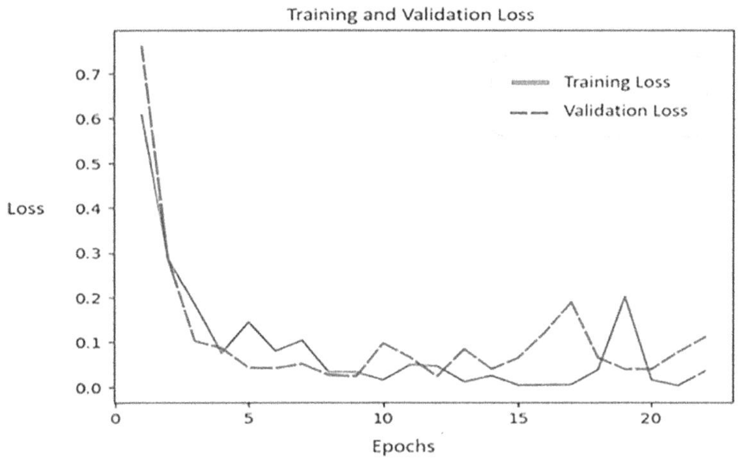

FIGURE 5.5 Training and Validation - Loss for epochs.

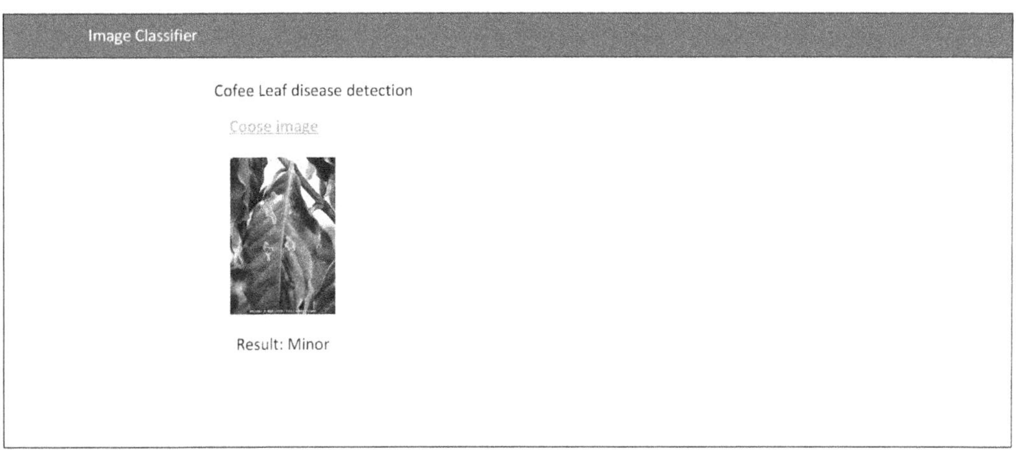

FIGURE 5.6 Web Interface of Disease Detection.

5.2 CONCLUSION

The combination of IoT and AI contributed to sustainable and efficient farming practices to define it as one of the methods for precision agriculture. However, the CNN-based coffee disease detection system demonstrates the power of IoT and AI to boost precision, sustainability, and farm productivity through widespread technologies by transforming and reshaping the agricultural landscape. However, as edge AI will speed up the process due to on-device processing, consuming less power and memory works well for the current phase when combined with 5G technologies.

REFERENCES

Anudeep, K. S., Prajwal, K. M., Sireesh Gowda, H., Venkatesh Prasad, G. N., & Jagadamba, G. (2024). Coffee leaf disease detection using convolution neural network. *Grenze International Journal of Engineering and Technology (GIJET)*, 10(2), 1314–1321.

Kamilaris, A., & Prenafeta-Boldú, F. X. (2018). Deep learning in agriculture: A survey. *Computers and Electronics in Agriculture*, 147, 70–90.

Kumar, P., Singh, A., & Roy, S. (2023). Smart irrigation systems powered by IoT and AI for sustainable agriculture. *Journal of Cleaner Production*, 421, 138542.

Malik, A., & Kumar, R. (2021). An overview on agriculture in India. *International Journal of Modern Agriculture*, 10(2), 2087–2095.

Mendeley Data. (n.d.). Datasets. https://data.mendeley.com/datasets

Mulla, D. J. (2013). Twenty-five years of remote sensing in precision agriculture: Key advances and remaining knowledge gaps. *Biosystems Engineering*, 114(4), 358–371.

Ravi, S., Mishra, D., & Tyagi, A. (2022). AI-enhanced crop modeling for smart farming systems. *AI in Agriculture*, 6, 32–45.

Sahu, A. K., Puhan, P. P., & Kumari, P. (2026). Artificial Intelligence-Driven Nanosensors for Plant Research and Agricultural Applications. *In Intelligent Nanotechnology in Plant Science and Agriculture*, 191–226, CRC Press.

Shamshiri, R. R., Kalantari, F., Ting, K. C., Thorp, K. R., Hameed, I. A., Weltzien, C., & Ismail, W. I. W. (2018). Advances in greenhouse automation and controlled environment agriculture: A transition to plant factories and urban agriculture. *International Journal of Agricultural and Biological Engineering*, 11(1), 1–22.

Singh, P., Mehta, R., & Zhang, Y. (2024). Use of deep learning for crop health monitoring through drone imagery. *IEEE Access*, 12, 33456–33468.

Wolfert, S., Ge, L., Verdouw, C., & Bogaardt, M. J. (2017). Big data in smart farming–A review. *Agricultural Systems*, 153, 69–80.

Xie, Y., Sha, Z., & Yu, M. (2008). Remote sensing imagery in vegetation mapping: A review. *Journal of Plant Ecology*, 1(1), 9–23. https://doi.org/10.1093/jpe/rtm005

Zhang, C., Liu, F., & Zhang, C. (2022). Artificial intelligence for sustainable agriculture: A review. *Agronomy*, 12(3), 738.

6 Cultivating the Future

Cutting-Edge Innovations for Sustainable and Precision Agriculture

Balakrishnan D, M. Parvathi, Akilandeeswari M, Deebika R, and P. Ravi

6.1 INTRODUCTION

With this rapid advancement and precision agriculture, automation, the Internet of Things, artificial intelligence, big data, and digital platforms, agriculture is witnessing a revolution. All of these new technologies are redesigning farm management by better resource productivity, safer food, and better environmental sustainability. It finally erased problems like climate change [1], water scarcity, and unsustainable agricultural practices that had accompanied human civilization for ages. The traditional ways must be bent towards data orientation and embracing the technologies within farm frontiers, facing the future projections of world populations to about 9.7 billion for the year 2050.

With the informatization drive in play, China leads a national momentum in driving agriculture in information technologies more through adaptation as complementary measures to benefit other sectors including the economic one [2]. China has computer science, mobile communication, sensor technology, embedded systems, and LAN communication to modernize its agriculture sector. All of these are attempts to seek one solution to an exceptionally sharp problem concerning inefficient management, overproduction pollution, and static returns on investment in the agriculture sector. This would be one of the infrastructural ways through which great benefits for both large-scale commercial farms and smallholder farmers will bridge the gap between the processes of urban and rural development.

The most important innovation in precision agriculture is the 5S precision agriculture system [3], which includes RS, GIS, GPS, ES, and SI. The system has been implemented with great success in geographically challenging regions of China's Taihang Mountains for optimized irrigation, land management, and crop yield enhancement. Precision agriculture enables farmers to monitor crop health, reduce input costs, and avoid environmental degradation through remote sensing, satellite imaging, and GIS-based spatial data analysis [4].

It forms the basis of smart agriculture, too. So, smart agriculture provides real-time monitoring and automation in this respect. IoT sensors, along with low-power wireless networks, RFID technology, and cloud-based analytics, give remote insights into the soil moisture level, livestock health, and conditions of crops [5]. One of the most prominent examples is IoT-driven aquaponics, where hydroponic or soil-less cultivation of plants accompanies aquaculture or fish farming. 6LoWPAN (IPv6 over Low-Power Wireless Personal Area Networks) and WSN automate nutrient circulation, water quality monitoring, and fish health management [6]. These technologies make the models of urban and vertical farming more efficient and sustainable.

All these technologies notwithstanding there are several barriers to the use of technology among small- and medium-scale farmers. Average prices of automated harvesters and drone sprayers are out of reach for most farmers, while AI farming robots can only be adopted through government

DOI: 10.1201/9781003613510-6

subsidies and policy interventions. This study examines the effects of subsidy-driven mechanization on the agricultural value chain using system dynamics modeling for the analysis of technology diffusion and economic growth. Studies indicate that financial incentives significantly increase technology adoption, replacing labor-intensive practices with smart, automated systems.

However, the penetration of technology is restricted by limited digital literacy and financial requirements. The study discusses the diffusion problem using complex network theory and simulates it on NetLogo. This study further elaborates on knowledge-sharing networks, training programs for farmers, and digital inclusion strategies that need to be put in place in this pursuit of mass adoption. To be adopted by many farmers, digital agricultural solutions must be friendly, scalable, and accessible.

Circular agricultural engineering is one of the most significant dimensions of sustainable agriculture. Such a form of agricultural engineering leads to efficient use of resources with minimal waste and consumption. The practice of circular agriculture considers recycling biowaste, closed-loop circulation of nutrients, and integrating renewable energy for higher fertility in the soil as well as reducing pollution and promoting ecologically friendly farming. The integrated farming models adopted by China as part of its circular economy policies have effectively minimized agricultural pollutants while optimizing sustainable resource utilization.

Big Data Analytics, along with AI-IoT, is therefore used by this agri so that it requires immediate monitoring in real-time and predictive analysis. These help assist farmers by showing measures regarding weather forecasts and even enable the detection in earlier stages of the breakout of diseases in crops. Traceability because blockchain introduces a high supply chain of high-quality food among people to trace its original source and certificates and storage.

As agriculture becomes increasingly data-intensive, autonomous, and connected, IoT, AI, and Big Data analytics are being integrated into self-managing smart farms. AI-powered drones enhance resource-efficient farming by automating irrigation in sensor-driven greenhouses. Real-time monitoring, predictive analytics, and precision farming will be further optimized through the deployment of 5G networks in agricultural landscapes.

However, the digital divide remains a major challenge, with developed regions benefiting the most. Barriers such as limited connectivity, high costs, and lack of technical knowledge hinder adoption in developing countries. This study explores how government policies and global partnerships can bridge this gap by ensuring accessible next-generation agricultural solutions.

Complying with the 2030 UN Sustainable Development Goals (SDGs), technology-driven farming is central to addressing food security, resource efficiency, and environmental sustainability. Cloud computing and big data analytics are transforming farm management as instant decisions are now made with real-time satellite information, weather stations, and smart sensor networks. Farm machinery and IoT-based systems assist growers in making quick decisions.

Precision viticulture is yet another flagship industry gaining momentum in the European Union, which holds nearly 45% of global vineyard ownership. Facilitating vineyard management systems through AI and IoT integration ensures optimal grape yield and sustainable wine economies. All such innovations also point towards the imperative of investing in cutting-edge farm technology that aims for the converging of the 2030 United Nations Sustainable Development Goals (SDGs) even more concretely towards ensured food security, climate resilience, and ethically utilizing resources.

The primary contributions of this research are the following. First, an extended system dynamics framework is created to investigate the role of subsidy-induced mechanization and technology extension in agriculture. Second, a simulation method based on complex network theory and the NetLogo platform is utilized to identify digital adoption barriers including cost, literacy, and infrastructure. Third, the study synthesizes empirical experience from a range of African, European, and Chinese cases to evaluate circular agriculture, precision agriculture, and aquaponics. Fourth, an IoT-AI-big data-blockchain combination framework is envisioned to provide advisory support for sustainable agricultural transformation. Lastly, actionable policy recommendations are presented to increase digital availability, farmer education, and inclusive access to smart agriculture technology in the direction of facilitating global sustainable development goals.

The rest of this chapter is structured as follows. Section 6.2 introduces the interdisciplinary research methodology of integration of empirical analysis, simulation modeling, and synthesis of case studies. Section 6.3 addresses the convergence of AI, big data, and blockchain technologies with mainstream agriculture. Section 6.4 addresses the concepts and practices of circular agricultural engineering for environmental sustainability. Section 6.5 addresses the technology architecture and deployment of precision agriculture, including the 5S strategy and IoT-based systems. Section 6.6 examines the effects of subsidy-led mechanization and assesses policy structures for embracing technology. Lastly, Section 6.7 concludes the study by posing major findings and future research areas for creating sturdy and data-sufficient farming communities.

6.2 METHODOLOGY

The research adopts an interdisciplinary and convergent research method toward new technology integration in agricultural settings based on principles of sustainability and precision agriculture. With the agricultural sector soon to be under a digital revolution, the convergence of emerging technologies like the Internet of Things (IoT) [7], artificial intelligence (AI), big data analytics, blockchain, automation, and circular agricultural engineering is pivotal in enhancing farm operations, resource use, and environmental impact reduction.

In order to construct an integrated image of such technology developments, this research applies a multi-dimensional research design with the fusion of qualitative and quantitative research approaches. These comprise empirical data analysis, simulation modeling, case studies, and conceptual models, providing a solid foundation for an assessment of the role of digital agriculture in productivity, food security, and climate resilience.

Since it's empirical data analysis-based, in the present research, estimates of the adoption of smart farm technology for agriculture, efficiency in the usage of resources, production improvement, and cost savings are provided. Simulation modeling provides an estimate of the long-term implications of smart farm technology, including forecasting adoption, scalability, and environmental footprint. Local and spatially differentiated farm circumstances come to the forefront using case studies, which use genuine examples of success with success stories and pitfalls. Furthermore, theoretical frameworks like network theory and system dynamics modeling allow for technology diffusion, economic viability, and policy intervention to be examined.

This research design enables strict analysis of how precision agriculture improves resource-use efficiency, agricultural productivity, and climate risk management. This research design is also contributing to the United Nations' Sustainable Development Goals (SDGs) since it ensures sustainable and productive agriculture [8]. Utilizing cross-disciplinary methods, the research will strive to formulate effective policy and strategic guidelines for submission to policymakers, researchers, and agricultural business leaders with actionable guidance to enable extensive adoption of future-ready agricultural technology towards building an efficient and resilient food system.

6.2.1 AI, Big Data, and Blockchain in Modern Agriculture

The integration of AI, big data, and blockchain is revolutionizing the contemporary agricultural scene by solving some of the most crucial decision-making, transparency, and supply chain efficiency problems. Algo-based platforms analyze enormous quantities of data gathered by IoT sensors, drones, satellites, and remote sensing devices to facilitate predictive modeling for disease detection, irrigation, and yield estimation [9]. Artificial intelligence-based decision-support systems facilitate precision agriculture through the analysis of historical data and real-time field data for maximum utilization of resources and risk avoidance [10]. Precision farming using artificial intelligence has been established to raise the yield of crops by 10–20% and reduce the cost of inputs by 15% wherever smart farming is adopted.

Big Data analytics further contributes to farm efficiency through the analysis of large data sets of soil conditions, weather patterns, and market patterns [11]. Predictive analytics identifies the optimum crop choice in accordance with regional factors like soil composition, precipitation, and temperature variation. Through studies, it has been observed that farmers practicing big data-driven decision-making have up to 25% more profitability in harvesting due to increased crop selection and adaptive cultivation strategies. In addition to this, Big Data enables supply chain management by monitoring farm produce from farm to market, minimizing post-harvest losses, and optimizing the efficiency of distribution.

Blockchain technology increases transparency and accountability in agri-food supply chains due to its decentralized and immutable ledger system. End-to-end tracking of produce becomes possible through it, and consumers can authenticate the origin, quality, and storage history of food items. Smart contracts enable automatic transactions with no intermediaries involved, ensuring fair trade practices and minimizing transactional overhead. Smart contracts allow automated transactions with no intermediaries involved, promoting equitable trade practices and reducing transactional overhead. These traceability platforms have been successfully implemented in organic farming systems, cutting fraudulent labeling by more than 30% and thus enhancing consumer confidence and trust in food quality and safety [13]. The combination of blockchain with sensor data from wireless sensor networks (WSNs) provides increased real-time visibility and monitoring of greenhouse operations, creating a strong architecture for smart farming environments. This system design is depicted in Figure 6.1, where the general architecture of a blockchain-based greenhouse monitoring system inspired from [14] is shown for how real-time sensor data can be integrated securely with blockchain systems for transparent agri-food management.

In spite of the optimism, these technologies are being held back due to scalability based on infrastructural constraints, quality of data concerns, and affordability concerns. Subsistence farmers lack computing power for the use of AI and Big Data, whereas the use of blockchain is brought back by prohibitively high power charges and integration charges. Case studies have illustrated that government policy, industrial collaborations, and subsidies can incentivize the adoption of such digital farm technologies. By analyzing AI-based applications of drones, blockchain-supported supply chain networks, and Big Data climate models, the research is interested in the extent of transformation that digital technologies are putting on modern farming.

6.2.2 Circular Agricultural Engineering for Sustainability

Circular agricultural engineering is a sustainability strategy by recycles waste, resource use, and renewable energy usage. The strategy, once more defending the circular economy strategy, prevents wastage and has increased effectiveness, combatting environmental degradation and sustainable agriculture. Compromising most are composting, biogas, integrated aquaponics, and recycling of nutrients, which promote maximum soil fertility and minimal pollution.

Waste-to-resource system forms the foundation of circular agriculture because it transforms agricultural waste into useful resources. Composting breaks down organic waste to transform it into fertilizer, reducing chemical inputs. Production of biogas captures the energy of decomposition as an alternative source of fossil fuel, and large aquaponics systems provide closed-loop systems for optimizing farm yield. Empirical evidence in China and Europe has shown that biogas utilization and composting can reduce fertilizer costs by 40% and restore 25% of soil organic content.

To measure economic and environmental sustainability in circular farming, the present work employs life cycle assessment (LCA). LCA enables the comparative assessment of resource productivity, carbon footprint, and net environmental footprint between conventional farming systems and circular farming schemes. Using organic composting, biogas generation, and closed-loop

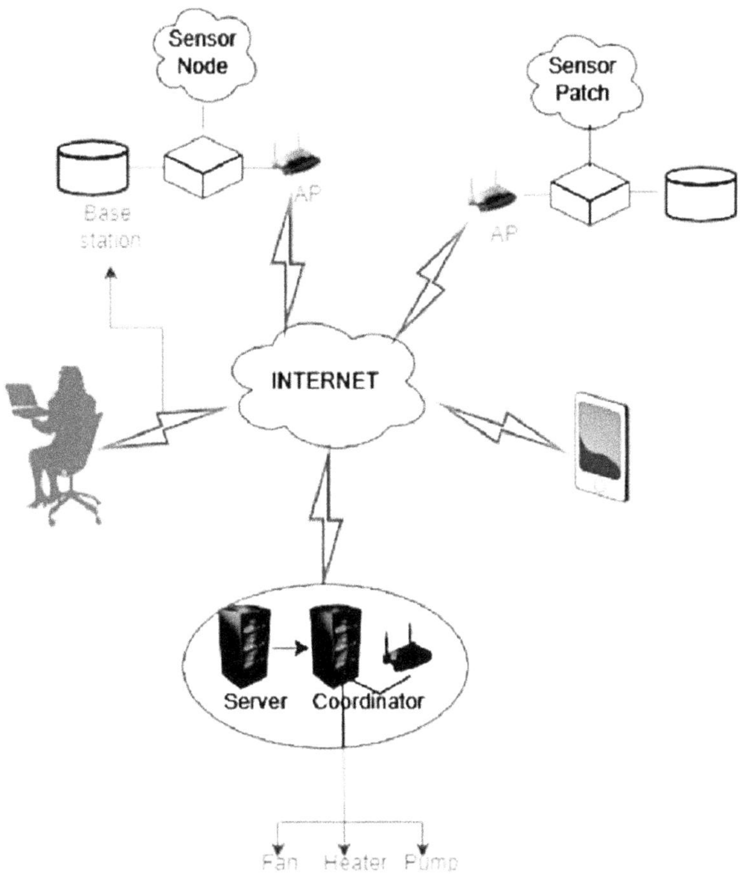

FIGURE 6.1 Overall architecture of WSN-based greenhouse [12].

irrigation numbers, appropriate measurements of farm-level efficiency and cost-effectiveness are further distilled in this study. These projects essentially contribute to mitigating GHG emissions, energy conservation, and water conservation. Figure 6.2 shows a working diagram of the design thinking framework with an illustration of the design thinking process for sustainable agriculture innovation and system-level integration.

Successful circular agriculture practice is proven by European, African, and Chinese case studies [15]. Europe offers an opportunity to set up advanced biowaste recycling and nutrient recovery technologies, whereby Chinese large-scale aquaponics farms utilize renewable energy technologies for upscaling. Biogas production to reduce energy shortages in Africa is gained through enhanced agricultural production. The flexibility of circular agriculture in various socioeconomic environments is illustrated through these case studies.

Simulation modeling suggests the long-term scalability of circular agriculture and offers a basis for take-up and scalability interpretations [16]. Empirical observation and predictive modeling are blended to situate drivers such as coordination of stakeholders, policy initiatives, and technological innovation. Network theory was used when examining knowledge-sharing networks among farming communities in rural areas and suggested that trajectories of effective farmer communication are responsible for shaping circular practice adoption structure.

Experimental trials also establish whether circular farm design is feasible, with biodegradable mulching film and aquaponics already practiced. They study how they affect overall farm efficiency,

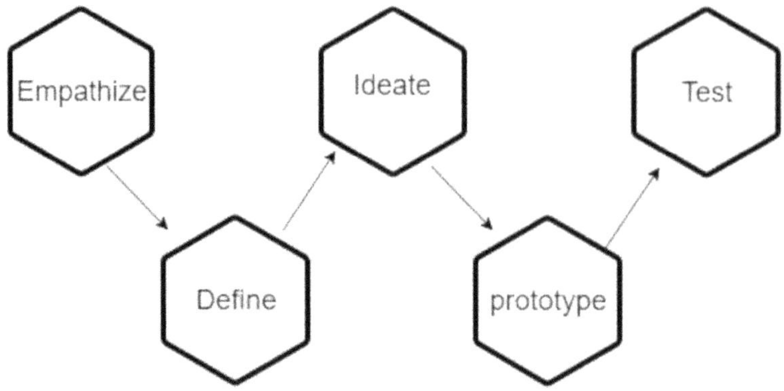

FIGURE 6.2 The design thinking methodology in the context of circular agricultural engineering, showing how empathize, define, ideate, prototype, and test phases advance sustainable agriculture innovation and system integration.

soil health, and crop yields. Biodegradable mulching film, for instance, reduces plastic use to optimal levels while conserving water to optimal levels through efficient soil cover, with real effects towards sustainable agriculture.

Agricultural scientists, environmental engineers, and sustainability experts are mentioned in the quest for more qualitative information on current applications of circular farm engineering. Experts mention the application of renewable energy, water recycling, and recovering nutrients in enabling sustainable agriculture. High investment cost, both technical capability requirements, and policy issues are mentioned as limitations, and therefore the need for all to collaborate and make circular farm technologies viable.

By increasing food security, decreasing the environmental impact, and effectively using resources, circular agricultural engineering strengthens the pursuit of sustainability globally, particularly the United Nations Sustainable Development Goals (SDGs). Harnessing renewable forms of energy, such as solar and wind energies, strengthens the environmental side of circular agriculture towards increasing sustainability and scalability.

6.2.3 Precision Agriculture Frameworks and Technologies: A Comprehensive Analysis

Precision agriculture (PA) is the new agricultural paradigm founded on emerging technologies such as the Internet of Things (IoT), artificial intelligence (AI), big data analytics, geospatial informatics, and automation. They have revolutionized farm management to a large extent through resource-efficient use, productivity, and environmental management.

With the onset of real-time monitoring technologies, remote sensing, GIS, and AI-based analytics, heretofore, precision agriculture had never been confronted by enigmas such as climatic uncertainty, input insecurity, and unsustainable production models. Consequently, the reconstituting blueprint for PA now seeks to emphasize data-enabled approaches with sensor networks and UAVs, along with predictive analytics, to enable decision-makers to effectively allocate resources instead of wasting them.

One of the more widely employed techniques in precision agriculture is the 5S system, which stands for Remote Sensing (RS), Geographic Information Systems (GIS), Global Positioning Systems (GPS), Expert Systems (ES), and System Integration (SI) [17]. This system supports overcoming geographical limitations namely, such as in the far-flung regions of China's Taihang Mountains-maximizing irrigation scheduling, land use, and crop yield to an optimized level. Using machine learning technologies and spatial data analysis, the 5S system allows for continuous soil

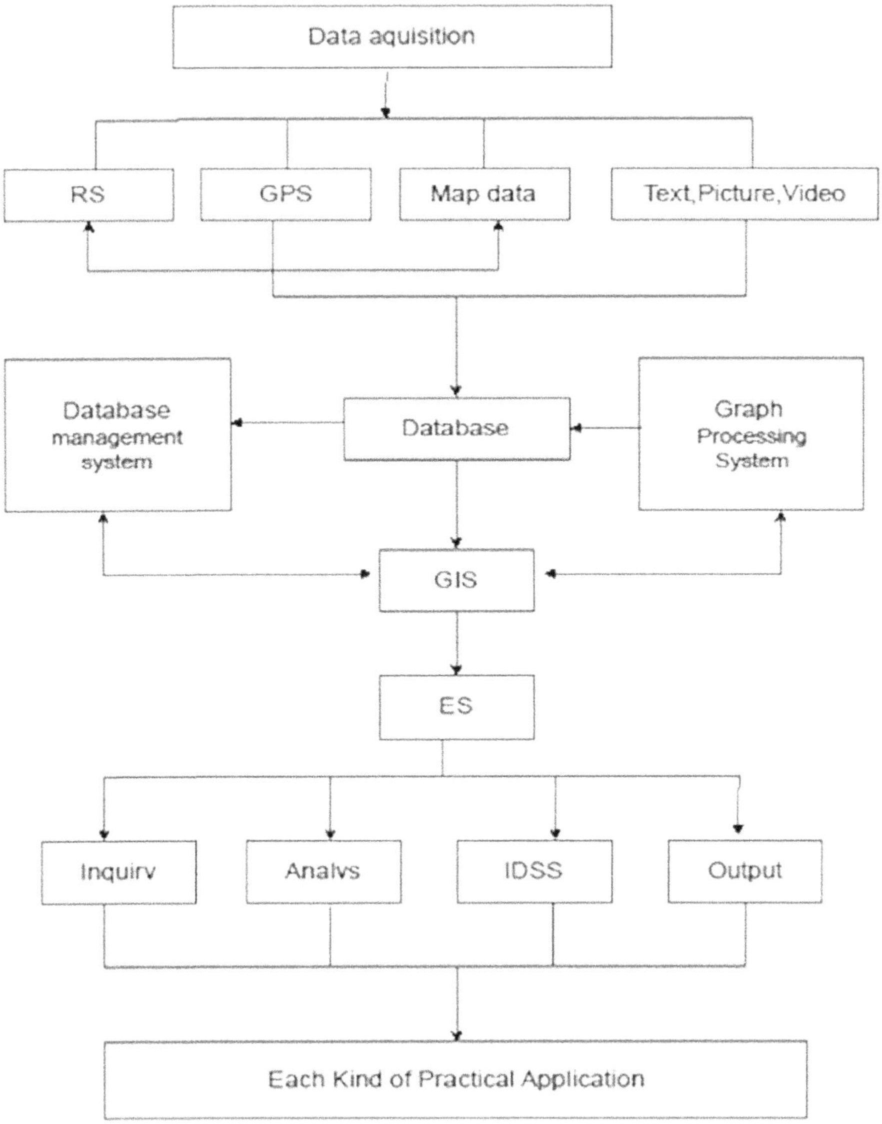

FIGURE 6.3 The 5S paradigm of precision agriculture, integrating RS, GIS, GPS, ES, and SI. By interconnecting the technologies, farmers can enhance monitoring, decision-making, and environmentally sound use of resources.

monitoring, early detection of pest and disease outbreaks, and effective usage of agricultural inputs for maximum sustainability and yield enhancement. In Figure 6.3, the operation and integration of these technologies are shown, accompanied by a schematic diagram of the 5S precision agriculture system.

IoT smart agriculture has also revolutionized the agriculture process with the ability to monitor crops, the content of soil moisture, and animal health in real time using low-power wireless communication, RFID, and cloud computing [18]. Concepts like aquaponics, wherein hydroponics and aquaculture are being integrated, reflect the extent of automation on the IoT-based platform for sustainable agriculture. These closed-loop nutrient recovery technologies have been of inestimable support to climate-resilient agriculture in the facilitation of proper water and nutrient management without any environmental cost. Adoption of PA technology is, however, skewed, especially for small farmers, with high capital outlays, low digital literacy, and limited infrastructure. Government action, in terms of offering financing subsidies and policy, has been crucial in precipitating the adoption of the technologies. System dynamics model-based research studies have examined the effects of subsidized mechanization on farm value chains and came to the view that economic incentives are the principal motive for supporting the shift away from labor-using and toward technology-using agriculture [19].

The use of Big Data analytics and artificial intelligence in farming these days cannot be overstated since they help the farmers foresee infestation by pests, development of disease, and production variations. Decision-support systems based on AI monitor an ocean of data from IoT sensors, satellite images, and meteorology to provide farmers with contextualized recommendations to facilitate smart farm decision-making. Farm supply chains have also been safeguarded by blockchain technology to render them traceable, authentic, and transparent throughout farm production. Blockchain smart contracts facilitate auto-transfers without middlemen and fair trade practices. Principles of Design Thinking have been used in crafting farmer-friendly digital solutions like VITIS, a groundbreaking precision viticulture platform for optimizing water and nutrition stress management for vineyards. This is the optimal demonstration of the potential of digital agriculture to couple traditional farming with a highly effective technology-based system.

Circular agricultural engineering, one of the most salient fields of sustainable precision agriculture, addresses the reuse of resources, waste reduction, and renewable energy application. Closed-loop nutrient recycling, organic waste recycling, and the creation of biogas to improve soil quality and avoid environmental degradation are concepts encouraged by the theoretical foundation of the circular economy. Circular agriculture policy, especially in China, has even further established the contribution of integrated approaches to agriculture in minimizing agricultural pollutants and achieving the highest level of resource efficiency. The use of the Life Cycle Assessment (LCA) method has also assisted in the overall analysis of the environmental and economic effects of circular agricultural systems. Circular and traditional agricultural system research shows phenomenal reduction of greenhouse gases, water, and energy consumption, which are the guarantees of circular agricultural systems towards global sustainability.

Despite all these developments, there are many hurdles to digital bridging between developed and developing nations. Limited connectivity, the expense of IoT devices, and the lack of adequate technical capabilities are some of the obstacles to slowing widescale uptake of PA technologies, especially in rural and resource-scarce regions. It is a plea to governments, technology innovators, and international institutions to join hands in providing the same degree of availability to digital agricultural solutions. Public–private partnership: targeted training courses and infra investment are needed in mass scale-up use of smart farm tech. The future of precision agriculture involves having networked autonomous farms, with drones equipped by making them intelligent, automated irrigation setups, and greenhouses lined by sensors. Roll-out of the 5G network in the regions will ignite real-time management of data, connectivity, and operational efficiency as a precursor for the high-productivity and green farming sector. With region-specific problem-solving and scalable solution designs, PA frameworks can make agriculture a data-friendly business that can solve increasing food needs for an increasingly large population of the world.

6.2.4 IoT and Automation in Smart Farming: Advancements and Challenges

The use of IoT and automation technology in agriculture today has revolutionized far less or indeed metamorphosed the old manner of farm labor into something by introducing highly effective data-based mechanisms inside it. Equipment based on IoT like GPS devices, soil probes, weather posts, and drones have simplified live tracking of the surroundings and crop health. The technologies provide data to cloud-based analysis systems in real time and are processed by artificial intelligence algorithms and machine learning models to optimize farm operations. Smart farming enhances productivity, sustainability, and profitability and reduces climate variability risks and inefficiencies in resource management through such digital technologies.

IoT smart agriculture technology has been a transformation as regards the optimization of resources and processes. Weather stations, soil sensors, and GPS-driven farm equipment have real-time information on temperature, moisture, pH of land, and plant health. Real-time data enable farmers to make appropriate decisions, making the best use of available resources without input wastage. Such computerized systems like computer watering utilize artificial intelligence-based reasoning for soil level water content-based watering control without disturbing the peak watering of crops. Computer-controlled tractors and artificially controlled pest management also ensure maximum farm activity is done in place of man-labor effort like plowing, sowing seeds, and spaying pesticide sprays. Precision agriculture drones complement smart agriculture with higher efficiency in fertilization and pest management, reduced chemical waste, and environmental balance.

Prototype farms in different agriculture systems have validated these technologies thoroughly, quantifying the impact of IoT-based automation on efficiency and sustainability for the farm. These findings offer a side-by-side comparison of performance measurements in terms of water, fertilizer use, labor return, and crop. Statistical inference such as regression and hypothesis testing have also been applied in the economic as well as environmental benefit through smart farming by unlocking the potential of IoT. All these have been the results of humongous savings of wastage of resources, crop resistance, and higher productivity. IoT technology in smart farming, for instance, facilitated 30% water savings by using AI-powered tractors and drones that cut the cost of labor and enhanced sowing and harvesting precision.

Emerging agri-tech hubs have shown scalable and economically sustainable models of intelligent farming in numerous countries such as Israel, Japan, and the Netherlands. For instance, precision agriculture in Israel has successfully set up an efficient method to save water in an arid region, while in the Netherlands, automation and environmental control are strategies to increase productivity in greenhouse systems. Because of these applications, IoT agriculture systems thus hold the potential to intervene major global issues like food insecurity, climate change, and resource depletion. Figure 6.4 suggests the conceptual design and working of either such intelligent agricultural systems or their architectural model depicting IoT integration in smart farming environments.

Scalability remains a hindrance for these systems with threats like initial costs, rural inaccessibility, and cyberattacks presenting themselves as barriers. Policies such as governmental subsidies, rural extension services, and specific programs supporting technology would probably find solutions for these challenges. With 5G, edge computing, and AI models coming online, agriculture is all set for a new working model of real-time, autonomous farm operations. In a nutshell, IoT and automation are shaping the agricultural landscape as a data-driven tech industry that can now respond to food demand variability from an ever-increasing global population.

In order to affirm disparate aspects of operational efficiency, environmental sustainability, and economic viability that grow behind smart farming technologies, we are willing to bring in more insights from pilot projects and global case studies alongside business analyses. The whole chain of real-time agricultural data from field sensors transiting off to a decision-making system is conceptualized in the form of Figure 6.5: a data visualization architecture diagram to complement further analytics and precision control in digital agriculture.

6.2.5 Subsidy-Driven Mechanization and Technology Diffusion: A Comprehensive Analysis

Agricultural mechanization is the most important thing for being productive, efficient, and sustainable in modern agriculture. However, most of the farmers are faced with financial, technical, and human resource limitations that restrain them from the use of mechanized processes. Government

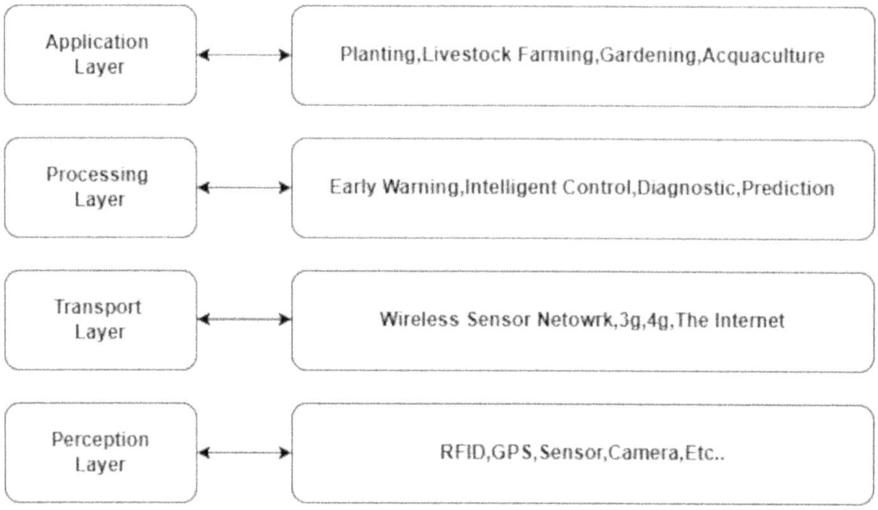

FIGURE 6.4 The convergence of IoT sensors, cloud infrastructure, and AI-based analytics in smart agriculture, and how sensor data is analyzed in real time and applied to manage automated farming machinery and maximize farming activities.

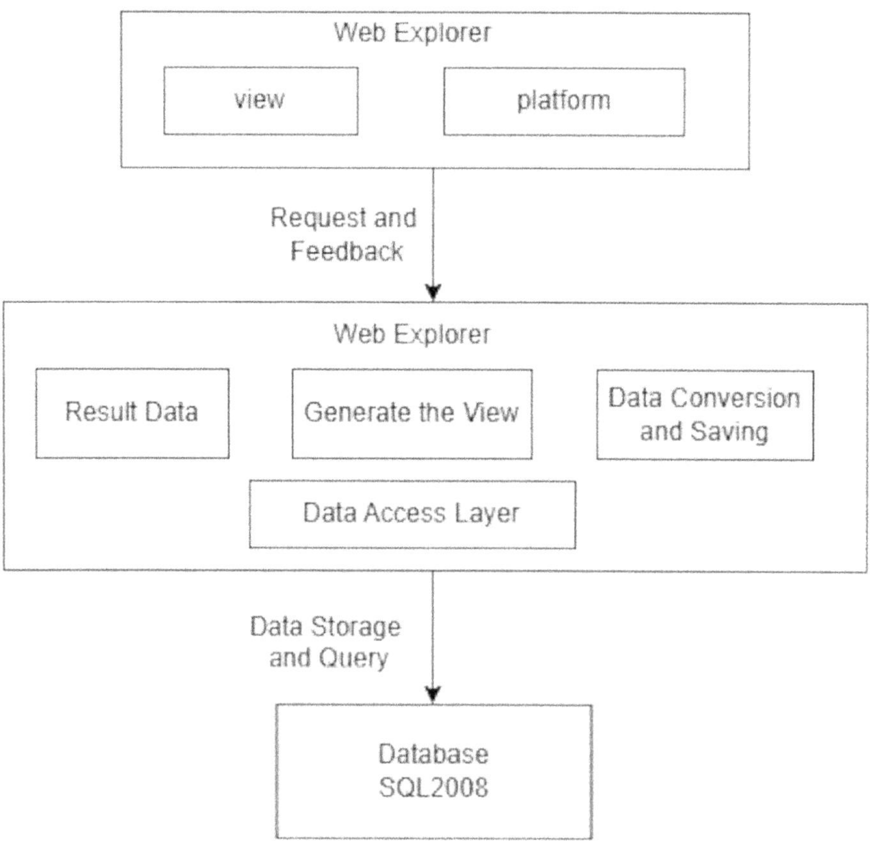

FIGURE 6.5 The agricultural data flow between field sensors and centralized analysis platforms, showing how visualization technologies enable decision-making, precision control, and monitoring of farm performance metrics.

subsidy, fiscal incentives, or cooperative finance models carry vast significance in breaking these hindrances, hence bringing about a shift in agriculture from hard work-based traditional approaches to technology-based and efficient agricultural practices [20]. In fact, the effect of such subsidies on mechanization is gaining more attention because a well-designed policy will increase technology adoption and make it economically viable as it becomes sustainable in the long term agricultural informatization. This study adopts a multi-disciplinary framework for exploring the role of subsidy-driven mechanization in transforming agriculture by combining economic analysis with computational modeling data analysis and case studies from globally.

One of the most daunting challenges is excessive up-front capital investment in equipment like tractors, mechanized irrigation, and precision agriculture equipment. This deters smallholder farmers from attempting to adopt these, as costs often become too much for them to handle. To testify to the efficiency of subsidies in overcoming financial constraints, a comparison will be made between farms that are financially assisted and those not receiving any financial assistance [21]. Key performance indicators of the analyzed financials, such as capital expenditure (CAPEX), operational costs (OPEX), return on investment (ROI), farm income, labor productivity, crop yields, and mechanization levels, will determine the impact of financial support on agricultural performance. Advanced statistical methods like structural equation modeling (SEM) and correlation analysis will be employed to assess the distribution levels in the context of subsidy dependency on technology adoption [22]. The findings showed that these subsidies increase farm productivity, financial viability, and mechanization mostly in states lacking adequate infrastructure. However, for the subsidies to be effective in the long run, they need to be context-specific to the agricultural needs because this should ensure that the financial assistance used is put to efficient use in promoting technological advancement.

This section particularly expounds on financial considerations but also socioeconomic and policy-related barriers affecting mechanization in agriculture. Through real-life farm-level case studies and interviews with stakeholders such as farmers, agronomists, policymakers, and agricultural companies, the reports indicate that key hurdles remain: huge capital outlays, lack of technical training, poor infrastructure, and broken-up landholdings. Farmer feedback indicates that there are still concerns on matters such as subsidy disbursement delays, poor access to cheap credit, and lack of good technical support services. The findings point out that a nexus of public agencies, private investor, and financiers need to pool their expertise, and design real and sustainable subsidy frameworks for agricultural modernization.

To explore the long-term effects of such interventions, this research utilizes an agent-based modeling (ABM) method to model different scenarios of subsidy allocation. The models analyze the dynamic interaction among subsidy levels, farmer adoption potential, and scalability of economic impacts over time. With training and support programs as complements, targeted subsidies can greatly hasten the process towards complete mechanization. But keeping the equilibrium between the initial financing and the establishment of self-propagating farmer techniques is essential to ensure long-term sustainability. The suggested computational architecture that directs this data-driven investigation is given in Figure 6.6, which is a schema for the Big Data and Machine Learning pipeline applied to subsidy optimization and policy modeling in smart agriculture.

For a truly global perspective, case studies of both success and failure are examined regarding subsidy-driven mechanization. The use of examples such as China's Smart Agriculture Initiative, the US Farm Bill Program, and India's PM-KISAN Scheme represents a few of the best cases of monetary intervention as it pertains to well-executed financial interventions. For example, China's initiative indicates how such an intelligent approach can integrate IoT and AI engineering with cooperative farming models to enhance productivity as well as sustainable resource management. Direct financial transfers complemented through policy frameworks make sure that all farmers can enjoy the benefits of such subsidies, forming the basis of India's PM-KISAN Scheme. Lastly, long planning combined with structured policy development has paid off within the US in the Farm

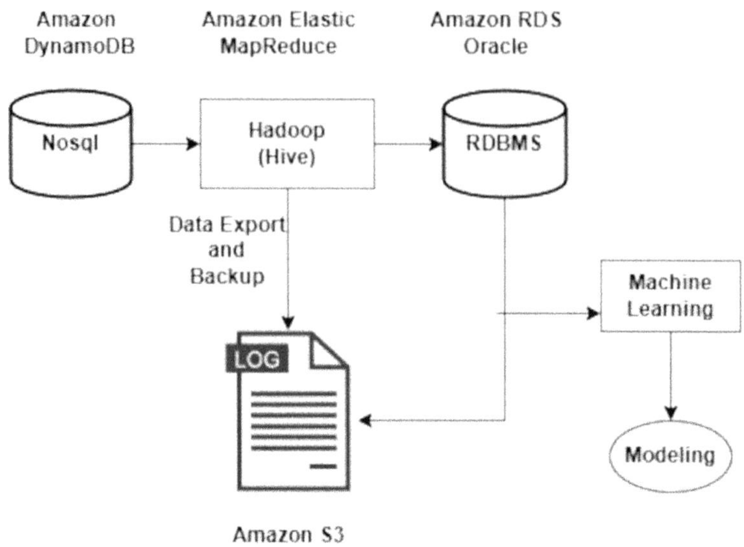

FIGURE 6.6 How farm, financial, and policy information are processed using AI and ML to enhance subsidy distribution and facilitate technology adoption in agriculture.

Bill Program, which has made provisions for both large-scale commercial farms and smallholder farmers. On the contrary, failed subsidy programs in various regions have most of their struggles in common, like poor fund allocation, transparency level, and low monitoring mechanisms, that have impeded the proper fulfillment of mechanization goals. This study's assessment of these available contrasted cases identifies best practices and lessons that can be used to craft better and more sustainable subsidy-based mechanization programs across the world.

The economic impacts of subsidy-driven mechanization are detailed with an economic impact study that compares the financial performance of subsidized farms with supporting government funds and non-subsidized farms, by such critical parameters as CAPEX, OPEX, ROI, labor charges, and revenue generation. The analyses have shown that farms benefiting from subsidies retain less financial burden and, thus, are in a position to invest in new and better machine technology and process improvements. Thus, improved operations result in better crop yields, lower demands on manual labor, and higher profits for farmers.

6.3 CONCLUSION AND FUTURE WORK

With automation and real-time data, farming is becoming more precise, less wasteful, and better equipped to handle the challenges of tomorrow. A typical application of this is the "5S" technology combining Remote Sensing (RS), GIS, GPS, Expert Systems (ES), and System Integration (SI) to enable farmers to optimize land use, irrigation, and health of the crop even on problematic terrain like the Taihang Mountains in China. Precision agriculture enabled farmers to use data in making decisions, closely monitor their crops with precise measurements using sensors, and automatize processes to avoid wastage and increase the harvest.

Smart farming with IoT takes it a step ahead by enabling round-the-clock monitoring of crop growth, soil health, and animal welfare. Cloud computing, RFID, and wireless sensor networks enable farmers to keep a check on their farms remotely and hence bring convenience and efficiency to the table. Aquaponics is a prime example where fish farming and hydroponic vegetable cultivation are done together. In this, IoT sensors monitor the water quality, nutrients, and fish health and have an equilibrial system that is self-sustaining. Such automation is revolutionary in the context of climate-resilient agriculture because it allows farmers to respond to shifting weather conditions and resource availability.

Government subsidies also have an important function in the transformation of agriculture through improved access to advanced technology for small to medium-scale farmers. Through subsidization support, agriculturalists can purchase high-tech farming equipment such as AI robots, drone sprayers, and mechanized harvesters to save on labor costs and enhance efficiency. The diffusion of agricultural technology remains uneven, nonetheless. Numerous agriculturalists are handicapped by illiteracy with digital matters, exorbitant prices, and poor infrastructure, which makes them incapable of getting access to such innovations in their entirety.

With technology leading agriculture into the intelligent age, the success formula is to make these innovations accessible to all farmers, no matter where they are or what equipment they possess. With capacity development programs, improved infrastructure, and affordable options, we can make the advantages of smart agriculture available to everyone and develop a more sustainable and resilient agriculture sector in the world. Studies employing advanced network theory and simulation models like NetLogo have demonstrated that training for farmers and knowledge networks are chief drivers of adoption levels of technology, ensuring that digital solutions can be scalable and accessible.

Circular farming has developed as the major strategy for sustainability, emphasizing minimal waste, conservation of the environment, and maximum utilization of resources [2]. By combining recycling of biowaste, closed-loop cycling of nutrients, and renewable power, circular agriculture diminishes pollution of the environment and improves the fertility of soils. China's policies for its circular economy are employed in case studies to discern how models for farms have minimized farm pollutants effectively and maximized the efficiency in resource utilization. Transition to regenerative, efficiency-based resource usage from traditional input–output agriculture connotes alignment with world sustainability targets to guarantee enduring food security.

Increased skepticism about climate change necessitates the use of big data analytics and artificial intelligence-driven predictive analytics to track agriculture in real time and forecast risks. Based on climate history, soil properties, and food market trends, AI-driven devices optimize yield estimates, improve effective crop rotation, and detect early indications of disease epidemics. Blockchain enhances traceability in agriculture by establishing transparency in food supply chains such that consumers can determine product origins, quality marks, and storage histories. Traceability systems with GIS mapping and ecological monitoring are shaping consumer confidence and motivating producers to maintain high-quality standards.

An example of such consumer-focused digital farming platforms is VITIS, an IoT precision viticulture platform designed to facilitate vineyard management through sensors, AI software, and remote sensing [23]. With better decision support for water and nutrient management, these platforms are an example of how user-friendly technology interfaces can help improve farmer take-up and facilitate optimal use of resources. The future of smart agriculture is even more data-focused with AI-fueled drone technology, robot-powered automatic irrigation systems, and sensor-based greenhouses to enable comprehensive autonomous agriculture systems [24]. The advent of 5G networks will further promote real-time connectivity, increase the efficiency of operations, and lead the way for AI-powered precision farming.

However, a significant digital divide is a concern as technologically superior agricultural regions lead developing countries in digital uptake. Denial of access to connectivity, affordability, and insufficient technical knowledge limits small farmers from effectively utilizing digital agriculture solutions. This gap needs to be bridged by governments, technology developers, and international institutions joining hands to make new agri-technologies affordable and inclusive. Specialized training programs, public–private partnerships, and investment in infrastructure will be essential to bridge the gap and achieve spectacular digital takeoff. The future of agriculture is about how technology can make it sustainable, efficient, and resilient.

The synergistic integration of IoT, AI, blockchain, and big data analytics with precision agriculture practices can be employed in line with the UN's SDGs so that food security, climate resilience, and optimal use of resources are ensured. As technology continues to transform farm operations, efforts should be taken to ensure there is equal access, scalability, and flexibility so that an environment-friendly, technology-driven agricultural system can be set up for the next generation.

REFERENCES

[1] J. Yin, Z. Yuan, D. Yan, Z. Yang, and Y. Wang, "Addressing climate change impacts on streamflow in the Jinsha River basin based on CMIP5 climate models," *Water*, vol. 10, no. 7, p. 910, 2018.

[2] D. Chasanidou, A. A. Gasparini, and E. Lee, "Design thinking methods and tools for innovation," in *Proc. Int. Conf. Design, User Experience, and Usability*, Cham: Springer, 2015, pp. 12–23.

[3] J. Bellvert, M. Mata, X. Vallverdú, C. Paris, and J. Marsal, "Optimizing precision irrigation of a vineyard to improve water use efficiency and profitability by using a decision-oriented vine water consumption model," *Precis. Agric.*, vol. 22, pp. 925–947, 2020.

[4] Y. Liu, Q. Chen, G. Liu, H. Liu, and Q. Yang, "EcoSense: A hardware approach to on-demand sensing in the Internet of Things," *IEEE Commun. Mag.*, vol. 54, no. 12, pp. 37–43, Dec. 2016.

[5] A. Kamilaris and A. Pitsillides, "Mobile phone computing and the Internet of Things: A survey," *IEEE Internet Things J.*, vol. 3, no. 6, pp. 885–898, Dec. 2016.

[6] K. Rong, G. Hu, Y. Lin, Y. Shi, and L. Guo, "Understanding business ecosystem using a 6C framework in Internet-of-Things-based sectors," *Int. J. Prod. Econ.*, vol. 159, pp. 41–55, Jan. 2015.

[7] D. P. Abreu, K. Velasquez, M. Curado, and E. Monteiro, "A resilient Internet of Things architecture for smart cities," *Ann. Telecommun.*, vol. 72, no. 12, pp. 19–30, 2017.

[8] A. Srivastava, "Technology assisted knowledge agriculture for sustainable development goals," *Adv. Crop Sci. Technol.*, vol. 6, no. 5, pp. 1–8, 2018.

[9] S. Yang, Y. Qiu, and B. Shi, "The key technology study on cloud computing platform for ECG monitoring based on regional Internet of Things," *Chin. J. Med. Instrum.*, vol. 40, no. 5, pp. 341–343, 2016.

[10] A. S. Elgazzar and A. S. Hegazi, "On complex adaptive systems and terrorism," *Phys. Lett. A*, vol. 337, pp. 127–129, Jan.–Feb. 2002.

[11] J. A. Tambo and T. Abdoulaye, "Climate change and agricultural technology adoption: The case of drought-tolerant maize in rural Nigeria," *J. Agric. Econ.*, vol. 17, pp. 277–292, Mar. 2012.

[12] J. A. Gutierrez, E. Callaway, and R. Barrett, *Low-Rate Wireless Personal Area Networks: Enabling Wireless Sensors with IEEE 802.15.4*, Piscataway, NJ, USA: IEEE Press, 2004.

[13] B. P. Kusumo Bintoro, "*NetLogo User Community Models: Twitter*," Jan. 15, 2013. [Online]. Available: http://ccl.northwestern.edu/netlogo/models/community/twitter.

[14] V. L. Narasimhan, A. A. Arvind, and K. Bever, "Greenhouse asset management using wireless sensor-actor networks," in *Proc. Int. Conf. Mobile Ubiquitous Comput. Syst. Serv. Technol.*, 2007, pp. 9–14.

[15] W. Yang, H. Liusheng, W. Junmin, and X. Hongli, "Wireless sensor networks for intensive irrigated agriculture," in *Proc. Int. Conf. Consum. Commun. Netw.*, Jan. 2007, pp. 197–201.

[16] A. Baggio, "Wireless sensor networks in precision agriculture," in *Proc. REALWSN'05,* Stockholm, Sweden, Jun. 20–21, 2005.

[17] Y. Y. Sheng and L. Z. Yu, "A study on the relationship between agricultural machinery buying and rural residents' income," *Stat. Decis.*, vol. 6, pp. 93–94, 2008.

[18] C. R. Gustafson, P. J. Barry, and S. T. Sonka, "Machinery investment decision: A simulated analysis for cash grain farms," *J. Agric. Econ.*, vol. 13, no. 2, pp. 244–253, 1988.

[19] X. Sun, M. Liqiang, and J. Zhang, "The comparative study of one- and two-dimensional agricultural products traceability," *J. Comput. CD Softw. Appl.*, vol. 12, pp. 57–58, 2014.

[20] Z. Zhengguo, "The application of GPS/GIS/GPRS in urban public transport system," *J. Technol.*, vol. 10, pp. 98–99, 2011.

[21] M. A. Peng, L. W. Qin, and Q. Q. Wang, *Introduction to Remote Sensing*, Beijing, China: Higher Education Press, 2001, p. 64.

[22] X. Ge, "Study on cropland information speediness collection and management system based on GPS and GIS," *M.S. thesis, Zhejiang Univ.,* 2003, p. 45.
[23] H. Mick, "Precision viticulture: A slow burn among smaller producers," *Aust. N. Z. Grapegrower Winemaker*, vol. 48, no. 672, pp. 34–38, 2020.
[24] X. He, Q. Ai, R. C. Qiu, W. Huang, L. Piao, and H. Liu, "A big data architecture design for smart grids based on random matrix theory," *IEEE Trans. Smart Grid.*, vol. 8, no. 2, pp. 674–686, Mar. 2017.

7 Convolutional Neural Networks for Potato Leaf Diseases Detection for Smart Agricultural Farming

Jagadamba G, Chayashree G, Hemavathi, and Varun Jayadeva

7.1 INTRODUCTION

Agriculture has undergone significant transformation, evolving from a booklet to a data-intensive science in the farms. In today's context, smart farming integrates Internet of Things (IoT) and Artificial Intelligence (AI) into agricultural practices to optimize operations and enhance productivity. In particular, smart farming utilizes image analysis techniques of deep learning. In this, the convolutional neural networks (CNNs) automatically identify and classify diseases from images of plant leaves, stems, or fruits. However, the CNN-based disease detection systems implemented through web applications. Farmers can detect early symptoms of crop diseases with high accuracy, receive real-time alerts and actionable recommendations, reduce reliance on manual inspections, and expert consultations and lastly apply targeted interventions (e.g., pesticide spraying) only where necessary, minimizing costs and environmental impact.

The AI-powered approach in today's smart agriculture enables farming decisions. The crop health monitoring can decrease yield losses (Anudeep et al., 2024). The AI has already transformed traditional methods into smart systems, real-time monitoring, and predictive decision-making, making smart agriculture an operative application to optimize agricultural inputs and outputs.

Traditionally, disease finding needs intricately trained agricultural experts performing visual inspections, which can be labor-intensive, subjective, and prone to error. The need for scalable, real time, and accurate diagnostic tools and systems has encouraged the adoption of artificial intelligence techniques in agriculture. Deep learning, particularly CNN, shows exceptional performance in image recognition and has yielded promising results in plant disease detection (Mohanty et al., 2016).

The potato, ranked behind rice, wheat, and maize production in India, is particularly vulnerable to many diseases. Remarkably, early blight caused by *Alternaria solani* and late blight caused by *Phytophthora infestans* have generally led to significant crop losses (Haverkort et al., 2009). These diseases not only reduce yield but also affect root product quality, resulting in economic losses for farmers. To address this issue, a case study was conducted to explore the adoption of AI-powered solutions for potato disease detection (Charanya et al., 2025).

This chapter is organized into six sections. A focus on introducing smart farming using deep learning techniques was done in the introduction section. Section 7.2 discusses the background work done on a CNN-based automated disease detection system. Section 7.3 discusses the methodology for integrating the image classification module into a user-friendly web interface. Section 7.4 outlines the proposed tool implementation, which empowers farmers and agronomists to make informed decisions early, thereby mitigating crop damage and enhancing productivity. Section 7.5

DOI: 10.1201/9781003613510-7

in the chapter includes the explored performance evaluation. Section 7.6 concludes with future possibilities and scope for improvement.

7.2 BACKGROUND AND RELATED WORKS

The incorporation of AI in agriculture has seen a significant rise in modern years, particularly for the detection of plant diseases using image classification techniques (Mohanty et al. 2016). Deep CNNs trained on the PlantVillage dataset to classify into 14 crop species across 26 diseases, achieving over 99% accuracy under controlled conditions. The work (Ferentinos, 2018) expanded on valuating deep learning models on a larger, multi-species dataset, confirming the generalizability of CNNs in plant pathology.

Sladojevic et al. (2016) demonstrated a CNN-based classifier for plant leaf diseases that achieved high accuracy across multiple crop types, but noted the need for large, well-labeled datasets and preprocessing techniques to reduce background noise. The (Amara et al., 2017) application of CNNs specifically to banana leaf diseases illustrates the adaptability of CNNs to specific crops and regions. However, these models are often not robust under real-world conditions, where leaf images may contain varied backgrounds and lighting conditions.

In the context of potato diseases, Barbedo et al. (2018) emphasized the importance of building crop-specific models and incorporating image augmentation to enhance model robustness. The current chapter builds on these insights by using a tailored CNN architecture and dataset balancing strategies to classify multiple classes of potato leaves, including early and late blight, and non-disease categories, within a unified framework.

The novelty of this study lies in its deployment focus, translating high-accuracy image classification into a practical, field-ready tool with real-time prediction capabilities for farmers, thus bridging the gap between research and application.

7.3 METHODOLOGY

7.3.1 System Overview of the General Disease Detection System

Figure 7.1 depicts the general potato disease detection system built with Streamlit for the image classification system. The interaction interface is facilitated through a "Browse Image" (upload) button. The uploaded image is then displayed to the user for verification purposes, ensuring the correct image has been selected. This step is crucial for preventing incorrect data from being processed, thereby improving the accuracy of the system's output. Furthermore, the proposed system offers an option to re-upload the image if the user realizes a mistake, ensuring flexibility and user control over the input process. However, this verification and reload process contributes significantly to the overall user experience and reliability of the system.

7.3.2 CNN Architecture

The prediction model architecture based on CNN is shown in Figure 7.2. Classify the image (Sanger Gopal, Rajasekar Velswamy, 2025) into the following categories:

- Early blight: If the image has early blight disease
- Healthy: If the image is of a healthy plant.
- Late blight: If the image has late blight disease.
- Non-Potato Leaf: If the image is a leaf but not a potato leaf.
- Object: If the image is an object or any non-leaf item.

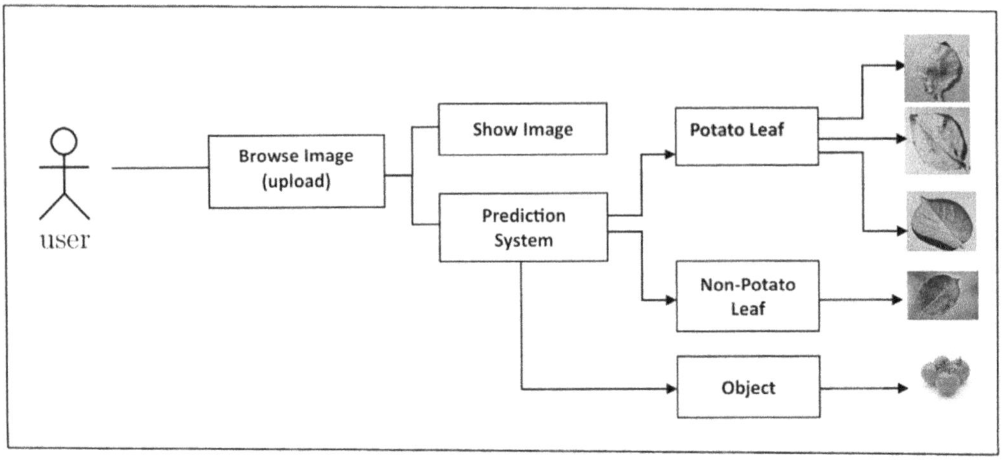

FIGURE 7.1 General potato disease detection system.

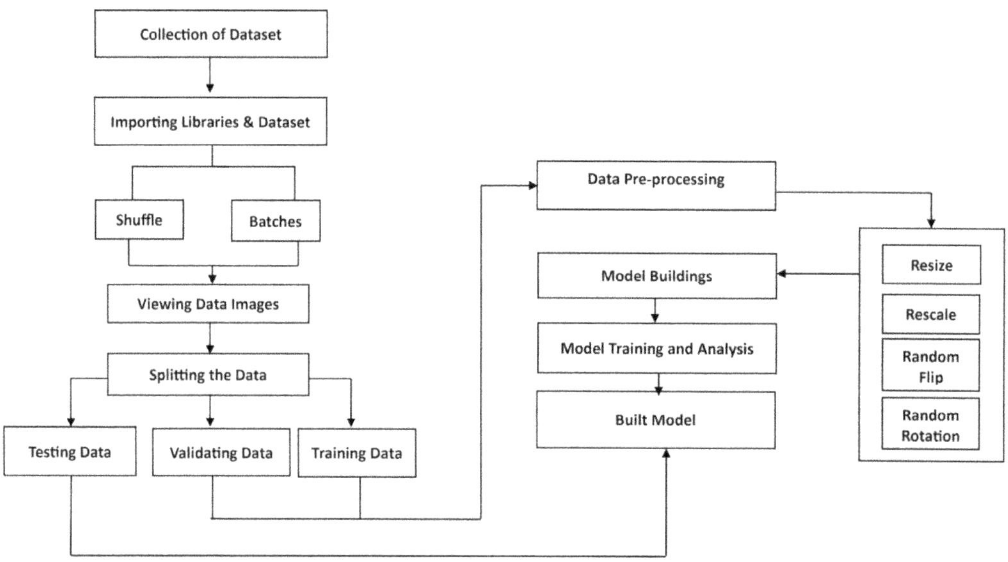

FIGURE 7.2 CNN-based potato disease prediction model.

7.3.3 Dataset

In a supervised machine learning project, data collection is the initial step, involving the organization of data into a structured format. In machine learning, a dataset is crucial as it provides the foundation for building and training models. Images were sourced from the publicly available PlantVillage dataset. The data were augmented using flipping, rotation, and shifting techniques to increase robustness. A sample dataset consisting of four different classes is shown in Figure 7.3.

Figure 7.3(a) defines a diseased potato leaf, Figure 7.3(b) defines the healthy potato leaf, Figure 7.3(c) defines a non-potato leaf, and Figure 7.3(d) defines an object. However, the dataset was split into 80-10-10 percentages for training, validation, and testing subsets.

FIGURE 7.3 Sample dataset.

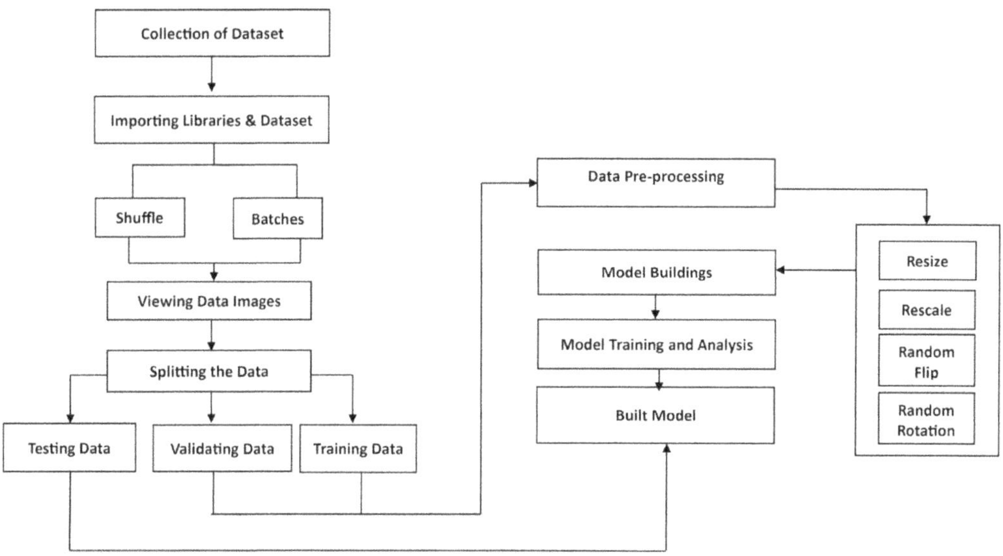

FIGURE 7.4 Flowchart of system implementation.

7.4 APPLICATION

Figure 7.4 represents a flowchart illustrating the procedural flow of the system. It provides a visual representation of the sequential steps involved, starting from the collection of datasets stage to the prediction stage.

Step 1: This includes the importing of Libraries and loading data like Pandas for effective data management, NumPy for numerical operations, and TensorFlow for constructing and training the CNN. Using tf.keras, reprocessing of the image dataset from the directory was done.

Step 2: This visualizes data samples to ensure data integrity and gain insights into the dataset's composition, visually inspecting a subset of the loaded images. This step aids in verifying label accuracy and understanding image characteristics like variations in orientation, lighting, and disease symptoms.

Step 3: This involves splitting the dataset into training and validation subsets.

Step 4: This performs data preprocessing to prepare images for effective training by resizing images to a uniform size. Additionally, normalization techniques are applied to standardize pixel values, ensuring consistent input across the network.
Step 5: This constructs the CNN Model using TensorFlow's Keras API. The architecture generally includes Conv2D for feature extraction and MaxPooling2D to decrease spatial dimensions and help prevent overfitting. Integration of dropout and batch normalization layers further enhances model stability. Accuracy metrics are employed to monitor training progress.
Step 6: This performs training and evaluation using CNN model.
Step 7: This performs the prediction of new images representing unseen data. This step involves checking the model's ability to become accustomed to unfamiliar data, highlighting its practical usefulness in real-world scenarios.

Thus, in the study, the training process involved 50 epochs by considering the Adam optimizer with the Sparse Categorical Crossentropy loss function.

7.5 EVALUATION AND RESULTS

The confusion matrix in Figure 7.5 showed minimal misclassification, with a high performance in distinguishing early and late blight diseases. The system demonstrated high reliability and low latency, making it appropriate for actual deployment.

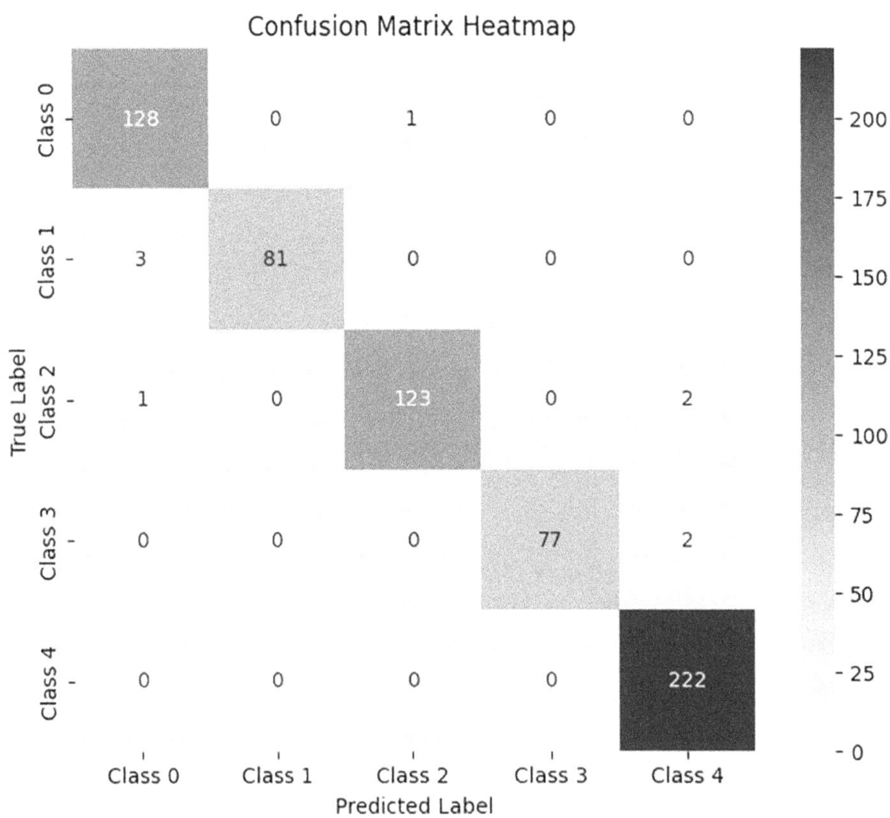

FIGURE 7.5 Confusion matrix of CNN model used in potato disease detection.

The study shows the model's learning process over epochs. Accuracy and loss curves for the CNN model used in potato disease detection are shown in Figures 7.6 and 7.7, respectively. The model was evaluated using F1 score, recall, and precision metrics. The performance of the same is shown in Figure 7.8. From the graph, out of all the instances predicted as diseased (or of a certain

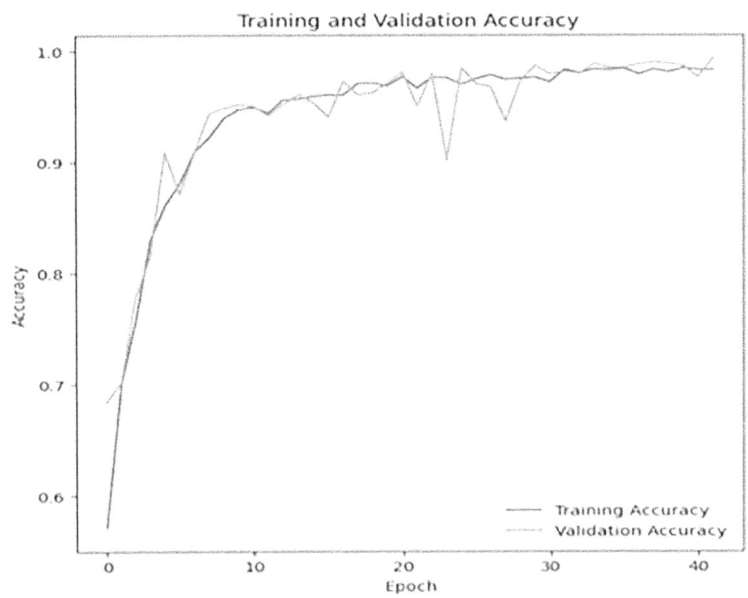

FIGURE 7.6 Training and validation accuracy curves for the CNN model used in potato disease detection.

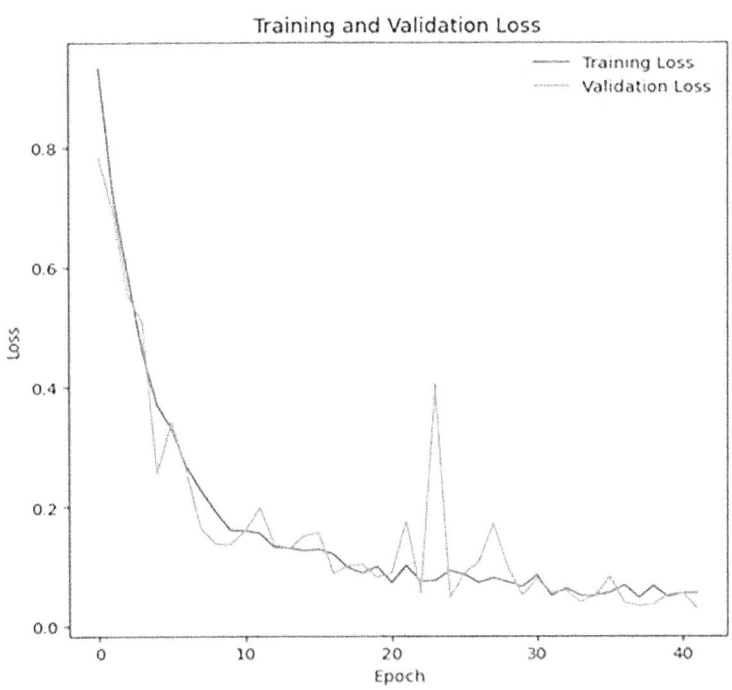

FIGURE 7.7 Training and validation loss curves for CNN model used in potato disease detection.

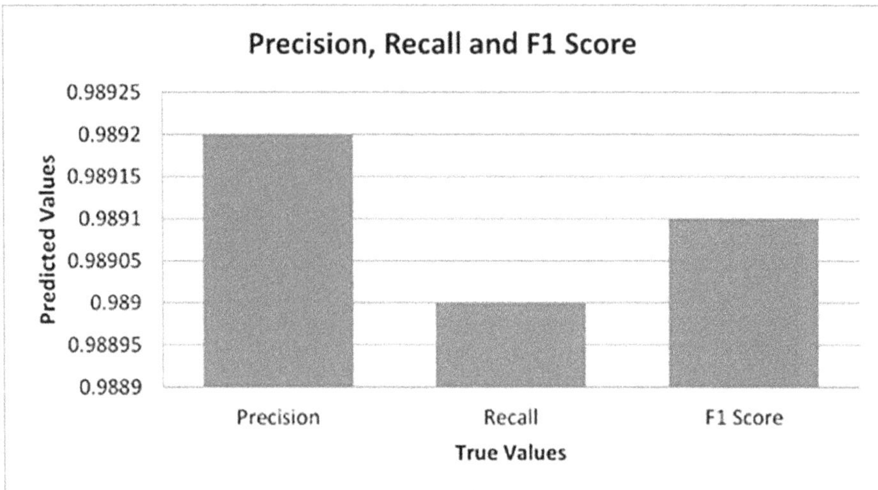

FIGURE 7.8 Performance metrics of the CNN-based potato leaf disease detection system.

disease), approximately 98.92% were found to be correct, i.e., the model is very good at avoiding false alarms (i.e., it rarely marks healthy potatoes as diseased). However, the model shows that it is slightly less effective in catching every diseased potato (it misses a small number of them), as the recall parameter is about 98.90. Lastly, the F1 Score balances the precision and recall tradeoff, as a high F1 score near 0.9891 means the model is well-balanced and overall performs very well in detection.

Figure 7.9 depicts the sample classifications with corresponding confidence values that are classified using the CNN model. For example, in Figure 7.9, the model correctly identified the leaf in the top-left Image as non-potato with 100% confidence. This illustrates the model's effectiveness in identifying non-potato leaves, reflecting its successful training. Such high accuracy and confidence are crucial for excluding non-potato leaves from disease analysis, enhancing efficiency, reducing errors, and highlighting its potential for real-world agricultural use. The category-wise values with evaluation parameters for different diseases identified by the model are given in Table 5.1.

7.7 CONCLUSION

This chapter demonstrated the effectiveness of a CNN-based system for the early detection of diseases in potato plants. By combining CNN with a user-friendly interface, the system offered scalable, precise, and practical solution for smart agriculture. It contributes to sustainable farming by enabling timely interventions, reducing crop losses, and improving food security.

Though the current system is limited to image-based classification, it can be extended with IoT to provide auxiliary data to enhance classification accuracy. Real-time alerts and dashboard visualization can further support decision-making for farmers.

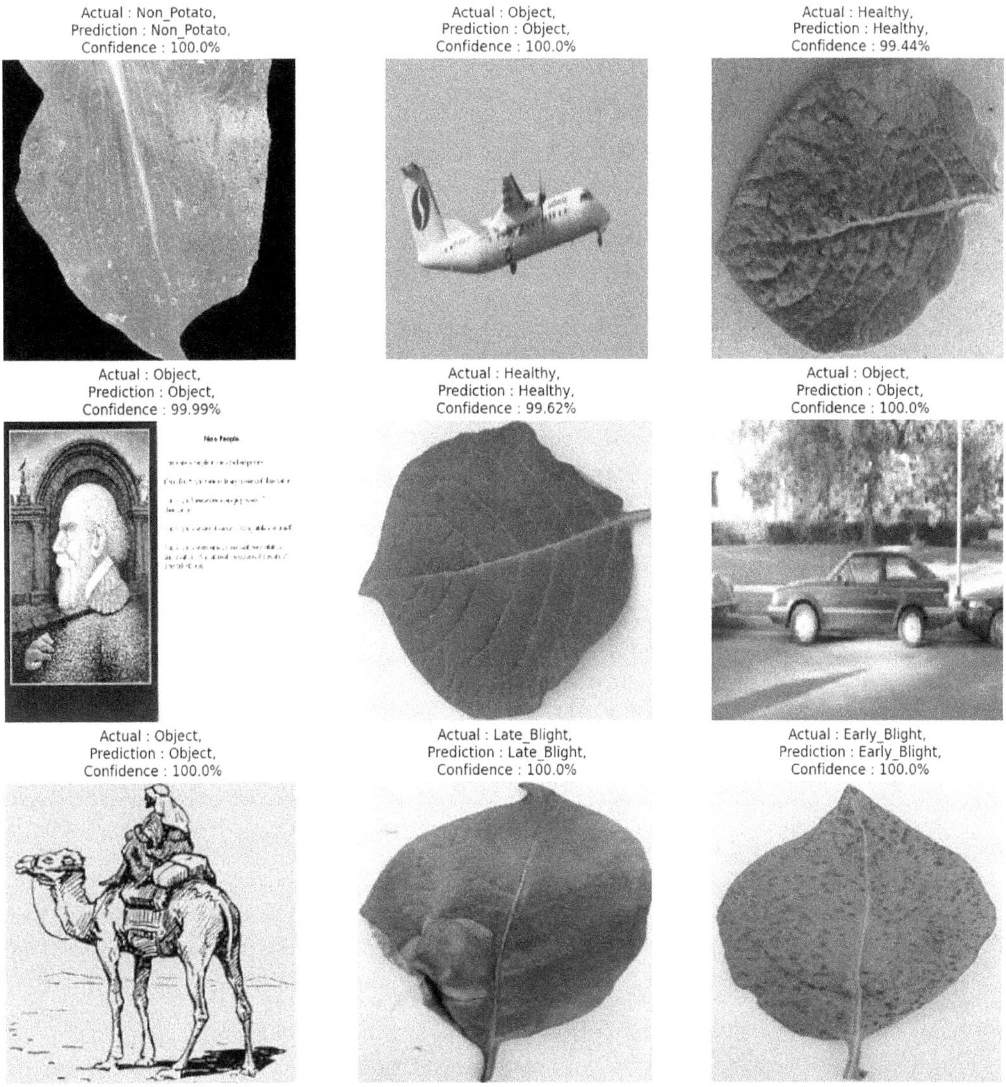

FIGURE 7.9 Sample classification results.

TABLE 7.1
Proposed Model's Performance Analysis

Classes	TP	FP	TN	FN	Precision	Recall	F1-Score
Object	222	4	413	0	0.9823	1.0000	1.0000
Early blight	128	4	507	1	0.9697	0.9922	0.9808
Healthy blight	81	0	555	3	1.0000	0.9643	0.9818
Late blight	123	1	511	2	0.9762	0.9840	0.9880
Non-potato leaf	77	0	558	2	0.9747	0.9747	0.9747

REFERENCES

Al-Adhaileh, M. H., A. Verma, T. H. H. Aldhyani, and D. Koundal (2023). Potato Blight Detection Using Fine-Tuned CNN, *Architecture*, 11(1516), Mathematics, MDPI. https://www.mdpi.com/2227-7390/11/6/1516

Amara, J., B. Bouaziz, and A. Algergawy (2017). A deep learning-based approach for banana leaf diseases classification. *In Datenbanksysteme für Business, Technologie und Web (BTW 2017)-Workshopband,* 79-88, Gesellschaft für Informatik eV.

Anudeep, K. S., K. M. Prajwal, H. Sireesh Gowda, G. N. Venkatesh Prasad, and G. Jagadamba (2024). Coffee Leaf Disease Detection Using Convolution Neural Network, *Grenze International Journal of Engineering and Technology (GIJET*), 10(2), 1314–1321.

Barbedo, J. G. (2018). Factors influencing the use of deep learning for plant disease recognition. *Biosystems engineering*, 172, 84-91.

Charanya, A., G. S. Gowthami, R. Mahesh, K. S. Yogananda, and G. Jagadamba (2025). Early Disease Detection in Potato Plant Using Deep Learning for Precise Agriculture, *GRENZE International Journal of Engineering and Technology (GIJET*), 11(1), 810.

Ferentinos, K. P. (2018). Deep learning models for plant disease detection and diagnosis. *Computers and electronics in agriculture*, 145, 311-318.

Gopal, S., and R. Velswamy (2025). Optimized Classification of Potato Leaf Disease Using EfficientNet-LITE and KE-SVM in Diverse Environments, *Frontiers in Plant Science*, 16, 1499909. ISSN=1664-462X

Gupta, A. K., K. Gupta, J. Jadhav, R. V. Deolekar, A. Nerurkar, and S. Deshpande (2019). Plant Disease Prediction Using Deep Learning and IoT, *6th International Conference on Computing for Sustainable Global Development (INDIACom)*, New Delhi, India.

Haverkort, A. J., P. C. Struik, R. G. F. Visser, and E. J. P. R. Jacobsen (2009). Applied biotechnology to combat late blight in potato caused by Phytophthora infestans. *Potato research*, 52(3), 249-264.

Kim, S., M. Lee, and C. Shin (2018). IoT-Based Strawberry Disease Prediction System for Smart Farming, *Sensors,* 8(4051).

Kingma, D. P., and J. L. Ba (2015). Adam: A Method for Stochastic Optimization, *3rd International Conference for Learning Representations*, San Diego.

Mohanty, S. P., D. P. Hughes, and M. Salathe (2016). Using deep learning for image-based plant disease detection. *Frontiers in plant science*, 7(1419).

Pinto, L. S., A. Ray, M. U. Reddy, P Perumal, and P. Aishwarya (2016). Crop Disease Classification Using Texture Analysis, *IEEE International Conference on Recent Trends in Electronics Information Communication Technology (RTEICT)*, Bangalore, India.

Sadiq, S., K. Razzaq Malik, W. Ali, and M. Munwar Iqbal (2023). Deep Learning-Based Disease Identification and Classification in Potato Leaves, *Journal of Computing & Biomedical Informatics*, 5(01), 13–25.

Sladojevic, S., M. Arsenovic, A. Anderla, D. Culibrk, and D. Stefanovic (2016). Deep neural networks based recognition of plant diseases by leaf image classification. *Computational intelligence and neuroscience*, 2016(1), 3289801.

Tiwari, D., M. Ashish, N. Gangwar, A. Sharma, S. Patel, and S. Bhardwaj (2020). Potato Leaf Diseases Detection Using Deep Learning, *2020 4th International Conference on Intelligent Computing and Control Systems (ICICCS)*, Madurai, India.

8 Sustainability and Environmental Impact of AI and IoT in Agriculture

Neroju Manasa, Guda Bhargavi, Kasula Vamshi Krishna, and Archana Kasula

8.1 INTRODUCTION

Technological impacts can still be positive or negative depending on the individual experiences. People can neither talk to each other nor make decisions without the help of machines. Artificial intelligence (AI) and Internet of Things (IoT) have advanced in the current world [1]. This is a collective form of interrelated devices that communicate with each other and perform actions based on information from various sensors using certain protocols. Smart irrigation systems are based on IoT sensors that measure the moisture level of the soil, and when it is necessary to irrigate the crops, water is automatically supplied in the required amount [2]. It has turned not longer to be considered as the business of the 21st century and it contributes between 17% and 18% of India's gross-domestic product (GDP) [3].

AI and IoT have a significant influence on modern farmers. Environment degradation as a result of modern farming techniques cannot be considered as a problem. The degradation of soil, water and air quality, primarily due to the excessive use of fertilisers and pesticides, poses a significant threat to biodiversity and ecological balance. Furthermore, the discharge of greenhouse gases in relation to agriculture has an impact on global warming, which is a change in climate that influences the living organisms and their environments. It is the process that sustainable agriculture stands out as an important approach that seeks to extract and use the various sources of production in a manner that does not harm the production of other natural resources and yet delivers a good yield and quality in production. For addressing these challenges, the integration of AI and IoT in agriculture has been evolved. AI, which originated in 1956, has today become integrated into several industries including agriculture. AI in agriculture combined with advancements in machine learning, neural networks and IoT seeks to simplify and optimise the farming procedures. This is not just a question of making things faster but of dealing with challenges like the weather, resources, and the financial exposures of farmers [4]. But the use of AI and IoT can make farmers relaxed by automating almost all the agricultural activities. Cyclical activities may happen to be planned from planting up to the harvesting stage. This means that the ability to predict demand and supply of agricultural products is now possible by using data analytics to help farmers financially.

This kind of integration is not without some challenges as these technologies seek to enter the agricultural sector. Some of the challenges include questions related to the economic relevance of AI systems and IoT, the necessity of specialised personnel and the difficulties of handling big data connected with agriculture. Besides, the sector has external challenges from climate change, destruction of environment and the socioeconomic aspects associated with rural constituencies.

Despite, there are great potential benefits that AI and IoT in agriculture can bring into reality. It provides water and soil management techniques, disease diagnosis in crops and precision farming

DOI: 10.1201/9781003613510-8

that make farming more effective with less harm to the environment, thus good returns for producers. The government realises these advantages and hence has employed various measures to encourage the use of these technologies in the agricultural sector [5].

Also, concerns about ethical issues including data protection, algorithmic profiling and the impact of AI utilisation in the labour market have to be considered. This chapter discusses the generality of the AI usage and suggests the development and implications of powerful computing models for sustainable agriculture and also aims at presenting an understanding how AI could be deployed in practice to address the issue of sustainable agriculture which in turn will address productivity challenges without harming the environment [6].

The most significant part of the method is the development of an AI–IoT framework tailored to address both issues of crops and the environment in sustainable agriculture. By using modern technology, including sensors, AI and data analysis, the approach enables specialists to decide on irrigation, fertilisers, and pest control. The goal is to use resources efficiently, use chemicals less and increase resilience to changes in the climate. Emphasising site-specific advice, this method allows small and marginal farmers to benefit from technology solutions that are affordable, usable and good for the environment.

8.2 ROLE OF AI AND IOT IN SUSTAINABLE AGRICULTURE

8.2.1 Precision Agriculture

AI models analyse satellite images and uses drones to monitor and analyse the field status, and soil and crop yields as well as disease detection. Through soil moisture sensors, weather station and nutrient monitoring devices, they gather data for the purpose of irrigation and fertilisation to avoid wastage of resources.

8.2.2 Smart Irrigation and Water Management

Smart irrigation and water management estimate the water demand depending on the climate conditions, the moisture content of the territory and the kind of crops worked. It adopts advanced and efficient water irrigation methods that maintain and control the water flow in order to enhance water usage and avoid wastage.

8.2.3 Pest and Disease Management

Warning signs of pest and diseases such as discoloration of plant tissues, wilting and die-back could be foreseen by drone and imaging AI analytics so as to apply pesticides at the right time. On this basis, the IoT device could be used for real-time monitoring and data analytics to notify farmers about pests' appearance and to prevent the non-adaptive usage of chemicals.

8.2.4 Soil Health and Nutrient Management

To cheque the use of fertiliser in cultivating the plants, they need to analyse the result of the soil test and suggest the right time and frequency as to when fertiliser could be applied in order to enrich the soil fertility. The controlled nutrient delivery systems control the input in relative to the actual soil status.

8.2.5 Climate-Smart Agriculture

Foresight of changes in climate acts as a warning system for catastrophic events, such as floods and droughts, hence allowing farmers to adjust to changes in climate conditions. IoT climate sensors are designed to cover the parameters of temperature, humidity and CO_2levels to ensure optimal climate for greenhouses and fields.

8.2.6 Supply Chain Optimisation

AI-driven market predictions aid farmers to determine the appropriate time to sell their crops depending on the price analysis. IoT logistics monitor storage conditions and transportation and therefore cuts down post-harvest loss.

8.2.7 Smart Farm Automation

AI robots assist gricultural operations such as weeding, harvesting and crop monitoring, thereby decreasing labour expenses. IoT M-based machinery sorts out the fuel consumption and general working efficiency of the machineries.

8.2.8 Waste Reduction and Circular Economy

Food waste prediction gives the necessary insights that aid in organising production and delivery in order to avoid wastage. Smart composting systems utilise IoT to track the process of decomposition of organic solid waste for efficient farming.

8.3 APPLICATIONS OF AI AND IOT IN SUSTAINABLE AGRICULTURE (FIGURE 8.1)

8.3.1 Crop and Soil Monitoring

Various crops and soils are monitored frequently by means of thermal or multispectral equipped robots or unmanned aerial vehicles. This makes the spray and control of watering with fertiliser easy. The sensors, therefore, measure the levels of various biomes that exist in the soils to ensure that the crops are highly nutritious. depending on the features of the land, AI defines which crops would be the most profitable.

8.3.2 Smart Greenhouses

IoT can contribute to increasing yield in smart greenhouses by enabling the creation of proportional controlling systems. They use sensors to provide an environment conducive for growth of crop that they cultivate. It is a remote-monitored system and the data processing involves the use of cloud servers [7]. Though, it minimises the usage of human intervention, the smart greenhouse regulates the amount of temperatures, light and even humidity in the surrounding.

8.3.3 Weather Monitoring

This connected devices of Internet of things to help in collection of actual weather and climate information. Farmers are in a position to analyse more effectively their crop needs because of proper projection [8]. Some system can also help the farmers to be alerted when the weather becomes too harsh so that they can protect the crops.

8.3.4 Drone System

Drones are used for both on ground and aerial ways to get the proper evaluation of crop health and monitor the insect infestation as well as to get a better view of the soil condition in less time. They could also be employed in monitoring, in real-time field data collection, seeds sowing, irrigation systems management as well as crop spraying [9].

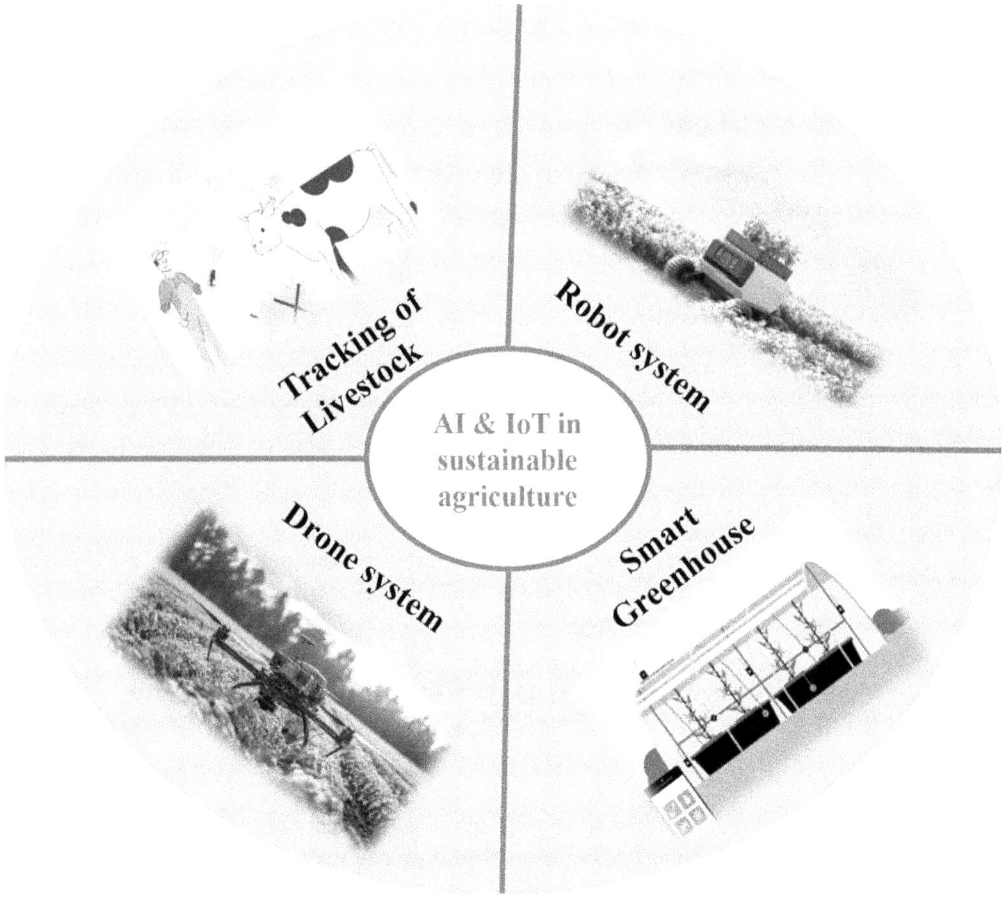

FIGURE 8.1 Applications of AI and IoT in sustainable agriculture. (Author.)

8.3.5 Robots

This cuts down the hiring of workers and amount of time required for their work since many functions on farms can be achieved by robots at once. They are used in crop monitoring and crop picking in better way than any manual practices can offer. They have obtained training in the use of AI, especially in helping to make proper care on crops to prevent the penetration of weeds. These devices are more effective in sorting the produces depending on their quality and packing them in a comparatively faster way than the conventional processes [10]. The application of Al in agriculture is aimed at helping the farmer to achieve greater production results and at the same time reduce the adverse effects on the environment.

8.3.6 Precision Farming

One of the widely used technologies in the agricultural sector is precision farming. This is the variable rate irrigation, soil moisture sampler and the water management for central through cloud service. As a result, the system takes advantage of the aspect of water resource through the use of sensors, self-driven machines and a network connection.

8.3.7 Tracking of Livestock

Through the use of wireless IoT network and related gadgets, it is possible to minimise the number of workforces needed at the ranch to monitor the cattle. IoT devices are capable of tracking the place of the animal and also its overall health [11]. On a large scale, farmers are able to handle livestock, and control disease by isolating the infected animals out of the herd. This also assists in the protection of the product and has the added benefit of maintaining low costs of cattle.

8.4 ADVANTAGES OF AI AND IOT IN AGRICULTURE

Applying AI and IoT has been of great benefit in the farming practices because of the improvement in efficiency and sustainability of farming. This means that AI is able to provide highly accurate data based on which proper decisions can be made to increase the samples to be produced as well as the efficient use of resources available. This information can best be used in forecasting proper planting dates, early disease detection on crops and efficient management of resources so as to promote better crop health. IoT complements AI as it entails the connection of devices that enables data collection on the environment and the state of crop growth at a given period of time. This means that the data is relevant in the management of resources, especially in agriculture. For example, the use of IoT sensor allows farmers to water the crops in the right manner, a factor crucial in water conservation and management. Overall, AI and IoT play a role in maintaining sustainable agricultural processes. They help in environment conservation by assisting farmers to efficiently applying fertilisers and pests against crops. Socially, one gets to appreciate the fact that technologies (Figure 8.2) in a way

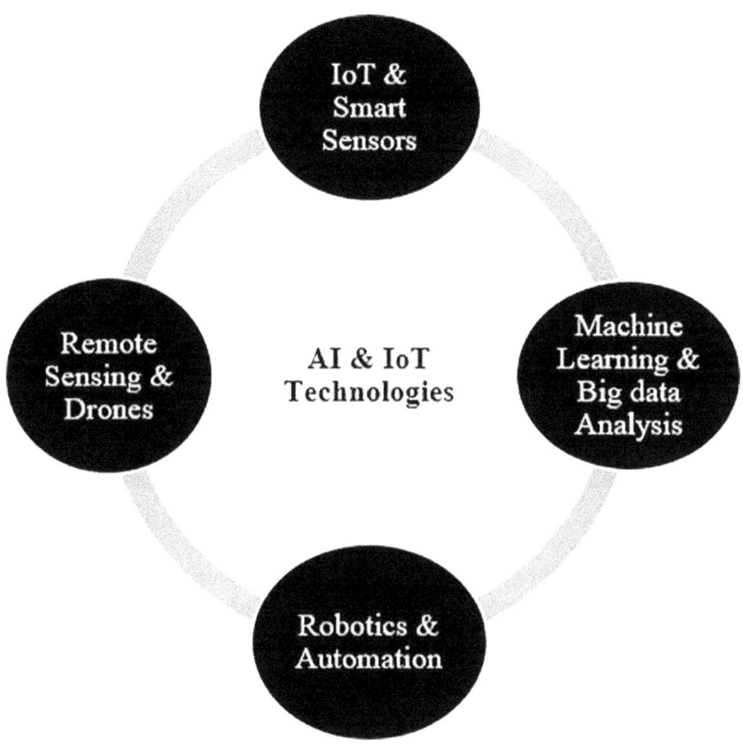

FIGURE 8.2 AI and IoT technologies. (Author.)

assist farmers to improve their incomes through lowering costs of business. Thus, the use of AI and IoT in agriculture is an innovative solution to feed the growing world population while conserving the environment. The use of AI and IoT has emerged as a revolution in lifting the technique of precision farming [12]. They dwell on how the cognitive computing in AI works through human decision-making to improve agricultural productivity. The study also cites the Microsoft Corporation that has implemented AI solutions in Andhra Pradesh, India, which has raised production by a remarkable 30% [13]. Multiple data incorporated by IoT are important for the core aspects in smart farming. It also explores how the radar proximity sensing and the remote sensing of soil are used for the precision analysis of the soil, and it also explains how robotics has been incorporated into data gathering for proper fertilisation. Moreover, there are advantages of imagery in farming based on the findings of drone imaging and computer vision. This can be used for identification of diseases and checking the readiness of the field for cultivation and management. The use of AI in irrigation automation and the substantial market potential for drone-based solutions in agriculture, particularly in precision farming and crop yield management, are also important [14].

8.5 IMPACT OF AI AND IOT IN AGRICULTURAL SECTOR

8.5.1 Potential Benefits

AI helps farmers deal with big data on different aspects such as climate conditions, soil and crop conditions, among others [15]. With the help of machine learning and predictive analytics, farmers are able to avoid planting times and control the irrigation and pests, thereby increasing crop productivity [16]. IoT indeed helps increase the productivity of farming by introducing various key processes such as sowing, watering and fertilisation with less or no human interference [17]. Another factor is that AI and IoT complement one another when it comes to using resources, for instance, water, fertilisers and pesticides. Like smart irrigation systems, IoT based systems use IoT sensors to measure the amount of moisture in the soil and when the crops require irrigation, water is supplied in the right quantity. Likewise, precision agriculture approaches based on AI assist in the use of fertilisers and pesticides for plantations based on the results of agricultural experiments and soil samples, which not only reduces costs, but also leads to less harm being done to the environment [18]. Real-time and historical data is used through AI-assisted decision support systems to assist the farmers in making decisions regarding their crops [19]. With satellite photography and maps, drones and AI farmers are able to diagnose the diseases affecting crops, conditions of the fields and the status of the plants. The smart devices in use that include weather sensors and automated machines help in data collection and analysis, enabling farmers to make effective decisions regarding planting, harvesting and supply chain management. Self-driven drones can patrol large areas of land that have crops to identify early signs of crop stress caused by diseases or pests. This is a proactive measure that assists in minimising cost and increasing crop yield in the farming sector. It has been established that the combination of AI and IoT in the farming industry leads to efficient use of fertilisers, water and land in farming. Modern innovative methods in farming intensification aim at increasing production and simultaneously reducing harm to the environment and effects of climate change, hence smart farming. AI usage and IoT in supply chain improve supply chain performance in terms of demand forecast, environmental conditions of products and general supply chain shipment. This chapter aims at exploring how AI technology in the area of forecasting assists farmers in predicting markets and IoT technology smart storage market assists in keeping perishable produce fresh, thus reducing post-harvest losses.

8.5.2 Sustainability

AI and IoT have improved the sustainability of farming practices through incorporation of smart technologies in farming. Application of AI in precision farming improves the usage of various resources like water, fertilisers and pesticides. Through data analytics, machine learning and IoT sensors deployment, farmers get the opportunity to control inputs. This makes the targeted approach not only efficient as far as resource use is concerned, excluding wastage, but it also impacts the environmental degradation resulting from agricultural practices [20]. In addition, the AI-applied models help in early detection of crop diseases and pest infections, determination of fertiliser and water requirements, and hence enhance resource utilisation [21]. Smart irrigation also relies on the IoT technology and allow controlling the water usage based on the current information about the state of soil. Likewise, in pest and diseases detection, AI helps in timely mitigating measures, hence averting over-reliance on chemical pesticides and providing environmentally friendly agriculture solutions [22]. In this way, innovative technologies are being implemented in the field of agriculture with the aim to increase food production, on the one hand, and reduce the effects of farming on natural environment, on the other hand.

8.5.3 Ethical Considerations

There are major benefits that arise from the implementation of AI technologies into agriculture, for example, higher crop yields, pest control and operational assets allocation. However, this change also brings considerable ethical issues into question. One such issue is algorithmic bias, which means that AI solution, based on specific algorithms and the data it has been trained on, can produce unfair or prejudiced results, which in the case of the presented problem may affect the fair distribution of resources between the farmers or the fairness of the decisions made for the crops [23]. In addition, the issue of responsibility comes into play whenever AI is to be embedded in decision-making that has a potential to affect the agricultural enterprise. Whenever AI results in an unfortunate occurrence like low yield or loss, the question arises as to whom it is appropriate to blame – the developers of the AI solutions, the farmers or other interested parties. This lack of clear accountability can endanger trust and/or the adoption of AI solutions to agriculture [24]. Also, as stated, transparency is an important factor. Many AI algorithms such as the machine learning algorithms work in such a way that their working cannot be easily explained to humans. This can lead to situations whereby the farmers, stakeholders, or regulators cannot understand the full picture or rationale of certain decisions, thus leading to a lack of full trust in these technologies [25]. Such concerns stipulate the need to have a proper ethical standpoint and legal methods of applying AI and IoT innovations in agriculture. These guidelines should be implemented to ensure that these technologies are used appropriately and are availed to all stakeholders in the agriculture sector in a manner that is most beneficial to them [26].

8.5.4 New Technologies

Precision farming is the latest revolution in agriculture as a result of technological intervention, especially through the use of AI and IoT. It is the hope to improve the farming activities, make the process unchallenging and in the long run to increase the productivity. This development involves using robotics in planting, harvesting and monitoring of the health of crops. This way it not only eliminates the expenses of hiring labour but also guarantees that the results concerning farming are accurate and standardised, thus increasing the yield and efficiency of the resources used [27]. One of the latest technologies in the advancement into the agriculture sector is the blockchain technology. Blockchain has most valuable properties for implementing secure, transparent and traceable systems and it is suitable when it comes to the supply chain management [28]. It also makes sure that all the flow from the farm to the market is recorded, which leads to prevention of quality compromise, food contamination and fraud. It also helps farmers to have improved market access since consumers and suppliers can easily identify the truth concerning agricultural products [29].

Furthermore, IoT technology is about the use of various types of sensors and other connected instruments that capture environmental factors including moisture content of the soil, temperature, humidity and health of the crops in real time [30]. This information enables farmers to get right decisions, undertake right watering pattern, right time for applying fertilisers and how the crop is likely to perform, thus helping apply the right measure of inputs required. As a result of consuming robotics, blockchain and IoT, AI in agriculture takes a new form of technology that turns farming into a modernised and intelligent process. These technologies remain useful in enhancing management of operations, tracking and marketing and productivity in future farming [18].

8.6 CHALLENGES IN ADOPTION

Despite the benefits of AI and IoT in agriculture, several barriers reduce their implementation in farming practises. Some of these factors include digital divide, where there is a discrepancy in the use of new technologies including the advanced ones in the developed world; this also applies to the peasant farmers being unable to take full advantage of the production innovation because they still lack adequate facilities in their rural settings. There are still problems of compatibility because the systems and standards between the AI tools, IoT devices and the agricultural machinery are not well integrated. In addition, data security and data privacy issues, which are not only critical but also cumbersome, are presented through the need to gather, store and analyse a large volume of agricultural data, in order to avoid misuse and unauthorised access or a breach. All these challenges need to be dealt with so that people can have an equal chance to benefit from AI in agriculture and also acknowledge responsible and ethical use of these techniques.

8.7 ENVIRONMENTAL IMPACT OF AI AND IOT IN AGRICULTURE

8.7.1 Positive Impacts

8.7.1.1 Resource Conservation

AI and IoT reduce wastage of water, fertilisers and pesticides as they are used in the right quantity and at the right time. The IoT sensors and AI can also track moisture, temperature and nutrient content of soils. This makes the use of water and other inputs to the crops to be efficient since the farmer can be in a position to apply water, fertilisers and even pesticides at the right time and in the required amounts. This optimisation is not only beneficial in terms of resources but also plays a role in resource conservation in the aspect of water, reduction of soil erosion and overuse of fertiliser and pesticides that are environmentally sensitive.

8.7.1.2 Greenhouse Gas Soil Emissions

The use of AI and IoT technologies help in cutting down the level of greenhouse gas emissions on farming practices. For instance, precision techniques driven by AI assist in the right application of fertilisers in a way that decreases the chances of nitrogen polluting the water, thereby reducing nitrous oxide emissions which are a form of greenhouse gas. They have helped in tracking or monitoring the livestock and their health in order to enhance their productivity as well as aiding in mitigation of methane gas. Also, the self-driving vehicles and drones using AI technology can effectively conduct the field duties such as sowing, watering or even pest control, hence reducing the use of fuel in powering machinery.

8.7.1.3 Conservation of Biodiversity and Ecosystem

Through proper utilisation of resources and investigating on the uses of enhanced chemicals AI and IoT can help conserve the ecosystem. The use of precise farming gives less contact with the environment and thus pesticide applications affecting positive insects and soil organisms, ecosystems, etc. Also, it assists in tracking the status of the soil and the breeds, which allows practices such as crop rotation, minimum tillage and Integrated Pest management (IPM) that would enhance ecosystem health.

8.7.2 Negative Impacts

8.7.2.1 Energy Consumption and Carbon Footprint of AI Systems

Despite the capacity to enhance agricultural practises, AI and IoT rely on consumption resources that may have a higher energy demand. AI and its parts such as machine learning, data processing, cloud computing, etc., require a comparatively high amount of energy in large-scale applications of such algorithms. The utilisation of AI in farming also has a carbon footprint in running these models, which if power is from sources such as coal might negate on the gain gotten. In the long-term perspective, there are two major problems for implementing AI, including the usage of renewable energy in AI systems and increasing energy efficiency of those systems.

8.7.2.2 Challenges of IoT Devices on E-Waste and Disposal

Through agricultural application of IoT devices, the world experiences congestion of e-waste. Devices involved in the IoT – for instance, sensors, cameras and automated machinery – will have a limited lifetime and may be a problem when discarded. This is because electronic is rich in heavy metals, which if not properly disposed can pollute the environment. The problem is to establish guidelines on how to effectively recycle the device or find new ways to reuse it, and how to design such devices to last longer to avoid being discarded.

8.8 CHALLENGES IN IMPLEMENTING AI AND IOT IN AGRICULTURE

Requiring Expertise: AI application and its integration in the agricultural sector require specialised knowledge, which can be acquired at all levels of education. It was also found that there is a discrepancy between farmers' knowledge and AI engineers' agricultural knowledge to facilitate better AI linkage suitable for the farmers [31].

Speed: Time plays a critical role when it comes to making choices regarding crop farming, the faster analysis of crop behaviour [32]. The machine learning model development takes time based on cropping data, which is usually collected annually or bi-annually [33].

Durability and Cost: AI technologies must be long-lasting and consequently cheap, especially in the developing countries where farmers practically operate on limited budgets. The expenses are not limited to the purchase of these sophisticated hardware as they also incur costs in terms of their maintenance.

Durable and Cheap: AI technologies should be long lasting and cheap, especially for the farmer who lacks capital to purchase the technologies. Maintenance of these highly developed basic devices also puts more financial pressure on the institution.

Continuous Improvement: The use of new and advanced technologies entails constant upgrading of machines and software, which raises issues on flexibility as well as installation.

Large Amount of Data: The high efficacy of an intelligent system depends on the quantity of data fed into the system. Handling big data is always a complex task.

Technical Proficiency Gap: Technical capability holds the farmers back; most of them do not possess adequate technical skills required to operate complex machinery.

Device Costs and Reliability: Specific to the target community, hardware costs and quality are the effective attributes especially in the development of devices and sensors used by small-scale farmers. While there are no critical requirements in terms of speed, the reliability of IoT systems is essential since their malfunction may lead to adverse effects on agriculture.

Security in IoT: In order to manage data in IoT networks properly many of which are constrained by way of processing and storage capacity. In IoT networks, data management is difficult due to the limited processing and storage capabilities of many IoT devices. Interoperability between these dissimilar devices is very important so that all devices would work well together.

Computational Framework: The existence of computing infrastructure defines how best data in all its forms can be processed. On the topic of cloud-based architectures, the computing and storage abilities may be sufficient as well.

Challenges Related to Big Data in the Framework of IoT: Due to large number of images and data from UAVs/satellites, the only way is to use data mining and AI. These must be techniques that can easily be used on large data sets as usually observed in the agricultural domain.

Speculation in Data Management: It is understandable that the complexity of the new data management techniques or practices, especially those practised under NoSQL, contributes to this uncertainty. It is critical to invest the correct data management tool since it depends on the application and performance needed.

8.9 FUTURE SCOPE OF RESEARCH

8.9.1 Development of Cost-Effective Technologies

The use of AI and IoT in farming can bring dramatic changes in the current practices but in terms of cost it is relatively high and thus strongly affects the large and mid-level farmers, not the marginal ones. Such interest is directed to investment in affordable and scalable systems based on AI and IoT. This includes developing almost invisible sensors for detecting the health of soil, water distribution to plantations and crop monitoring. Also, attempts are ongoing to develop end-user applications with integrated data analytics at low cost that will help farmers in arriving at new decisions without the use of complex tools [34].

8.9.2 Advanced AI Models for Diverse Crops

Most of the models of AI today are designed to support basic and commercial crops and there is a lack of technologies that support less used crops. It is therefore important that AI should be applied to many crops as possible, including the indigenous and underutilised crops, to help improve the productivity of all the sectors. Thus, research should be aimed at training AI models utilising data from the given region to offer reliable advices on soil condition, disease resistance and favourable conditions for growth [35].

8.9.3 A Sample Solution to Improve IoT Connectivity in Rural Areas

Among the challenges that forestall the adoption of IoT in the demand side, scarcity of Internet connection especially in the remote and rural agriculture-producing regions stands out. Hence, continued study for connectivity solutions like LPWAN, satellite internet services, and edge computing will help in boosting the IoT application possibilities in these regions. By ensuring that such IoT-based smart farming solutions can leverage a cost-effective and sustainable network foundation, these solutions can be harmonised into rural agriculture systems for productivity and resource optimisation [36].

8.9.4 Integration of AI with Traditional Farming Knowledge

The advancement of precision agriculture through the use of AI is ongoing but farmers' knowledge acquired through experience cannot be dismissed. It is possible to combine the use of AI with the traditional forms of farming to come up with versatile practises which are efficient in terms of production as well as sustainable. Using the trend data, indigenous wisdom and traditional agriculture practises, it is possible to train the recommendation system to be more bidirectional and aligned with local farming practices while following technological advancement [37].

8.9.5 AI and IoT in Climate Change Adaptation

Agriculture produces are at risk by increased climate change whereby the climatic change results due to increased incidences in harsh weather, droughts and extreme temperatures, which reduces crop yields. AI climate probability models can be employed and studied and this can help farmers to be able to detect and avert these challenges. AI and IoT can be used to monitor the moisture, temperature and humidity content in the soil, thus improving climate smart methods of farming such as growing crops that are resistant to drought, managing irrigation practices as well as management of fertiliser use. Finding ways to create AI models that can forecast climate changes, recommendations that would increase crop resistance can be of immense value [38].

8.9.6 Data Privacy and Security in Agri-Tech

Given that both AI and IoT require the accumulation of vast amounts of data, data protection and confidentiality are more important. When it comes to farm information such as the fertility levels of the soils, crop productivity or market prices and trends, technology results in storage and transfer of the information through the Internet. Therefore, research must work towards designing secure data security mechanisms. Other measures that can be adopted to guard the farmer data from exploitative use while avoiding restrictions to genuine utilisation of agriculture analysis include the use of permissioned distributed models.

8.9.7 Integration of Blockchain for Transparency and Traceability

This is in view of the fact that the application of blockchain technology will help introduce the concepts of transparency and traceability in the agricultural supply chain. Blockchain combined with AI and IoT in the field of research can help in the record keeping of crop production, supply chain hosting and food safety issues involved in the transactions. Smart contracts based on the blockchain protect farmers from unfair or exploitative sales in trades from the farms; decrease fraud in agriculture trade and bring about the traceability of food items from the farm to the consumer, thereby enhancing the agriculture trade, security and quality of foods that reach the consumer's table.

8.9.8 Autonomous Machinery and Robotics in Agriculture

Autonomous machinery and robots are giving a new look to the upcoming agriculture, decreasing the employment cost and increasing the working efficiency. In this domain, research is conducted towards the creation of AI-based robots for sowing seeds, eradicating weeds, harvesting, and identifying and eliminating pests without outside help.

8.9.9 Customised AI Solutions for Different Crops and Geographical Conditions

Farming or cultivation as it is widely practised is not universal since it is influenced by factors such as climate, type of soil and the types of crops to be grown. There is evidence that a universal pack does not always work all the time. This brings research on the development of specific model that works for a given geographical region or given crops. AI through geographic data and data analysis in machines can offer recommendations on matters such as fertility management, irrigation and disease control, thus improving farm productivity.

8.9.10 AI in Pest and Disease Prediction and Management

Pests and diseases are still a major problem for the farmer and crop losses around the world. The integration of computer vision and deep learning for pest and disease identification in real-time plant health analysis to warn farmers is necessary. Such research can result in the improvement of AI capacity in developing pest management systems, modelling of diseases and pest, and precise application of pest control agents – thus saving crop yields and preventing harm to the environment [39].

8.10 CONCLUSION

Combining AI and IoT in agriculture can help increase sustainability, efficiency and productivity. Farming methods that have been practised for many years often find it challenging to use resources wisely and protect the environment. AI and IoT, farmers can practise precision agriculture with constant monitoring, detailed analysis and automation, so that they can harvest more and manage their supply chain better. However, these new technologies have not yet been widely accepted because of a range of problems due to the digital gap, inability for systems to share information effectively, worries about privacy and security, and the demand for experienced staff. To address these problems, governments, researchers, developers and agricultural contributors need to cooperate and create inexpensive and effective solutions. It is just as crucial to add transparency and accountability to these systems to make sure they are used ethically and equally.

Further work should aim at making AI and IoT solutions accessible and budget-friendly for rural and remote regions, as well as combining them with traditional farming methods. It is crucial to pay special attention to data protection as well as designing ethical frameworks. Besides, more research should be conducted on employing blockchain to increase supply chain visibility and self-operating machines in farming. When overcome, these challenges allow the sector to make full use of AI and IoT for sustainable farming, good environmental outcomes and enduring stability.

REFERENCES

1. Joseph, Ferdin Joe John. "A review of IoT implementations in environment and agriculture." *JARES* 6, no. 1&2 (2019): 1–5.
2. Sivaganesan, D. "Design and development AI-enabled edge computing for intelligent-IoT applications." *Journal of trends in Computer Science and Smart technology (TCSST)* 1, no. 02 (2019): 84–94.
3. Sinwar, Deepak, Vijaypal Singh Dhaka, Manoj Kumar Sharma, and Geeta Rani. "AI-based yield prediction and smart irrigation." *Internet of Things and Analytics for Agriculture* 2 (2020): 155–180.
4. Ahmad, Latief, and Firasath Nabi. *Agriculture 5.0: Artificial intelligence, IoT and machine learning*. CRC Press, 2021.
5. Bhat, Showkat Ahmad, and Nen-Fu Huang. "Big data and AI revolution in precision agriculture: Survey and challenges." *Ieee Access* 9 (2021): 110209–110222.
6. Siau, Keng L., and Yin Yang. "Impact of artificial intelligence, robotics, and machine learning on sales and marketing." (2017).
7. Gai, Keke, Meikang Qiu, Hui Zhao, Lixin Tao, and Ziliang Zong. "Dynamic energy-aware cloudlet-based mobile cloud computing model for green computing." *Journal of Network and Computer Applications* 59 (2016): 46–54.
8. Kale, Shivani S., and Preeti S. Patil. "Data mining technology with fuzzy logic, neural networks and machine learning for agriculture." In *Data Management, Analytics and Innovation: Proceedings of ICDMAI 2018, Volume 2*, pp. 79–87. Springer Singapore, 2019.
9. Kamlesh Lakhwani, Gianey Hemant, Agarwal Niket, and Gupta Shashank. "Development of IoT for smart agriculture a review." In *Emerging Trends in Expert Applications and Security: Proceedings of ICETEAS 2018*, pp. 425–432. Springer Singapore, 2019.
10. Rajeswari, S. K. R. K. A., Kannan Suthendran, and K. Rajakumar. "A smart agricultural model by integrating IoT, mobile and cloud-based big data analytics." In *2017 International Conference on Intelligent Computing and Control (I2C2)*, pp. 1–5. IEEE, 2017.

11. Raja, S. P., T. Dhiliphan Rajkumar, and Vivek Pandiya Raj. "Internet of Things: Challenges, issues and applications." *Journal of Circuits, Systems and Computers* 27, no. 12 (2018): 1830007.
12. Ndiema, Kiptoch William. "Implications of artificial intelligence (AI) in land management." *Journal of Computer Science and Technology (JCST)* 2, no. 1 (2024): 1–9.
13. Chandra, Vinod SS. "Role of artificial intelligence in Indian agriculture: A review." *Agricultural Reviews* 44, no. 4 (2023): 558–562.
14. Khandelwal, Paras M., and Himanshu Chavhan. "Artificial intelligence in agriculture: An emerging era of research." *Research Gate Publication* 1, no. 1 (2019): 56–60.
15. Saha, Sanchita, Ashok Ghimire, Mia Md Tofayel Gonee Manik, Anamika Tiwari, and Md Ahsan Ullah Imran. "Exploring benefits, overcoming challenges, and shaping future trends of artificial intelligence application in agricultural industry." *The American Journal of Agriculture and Biomedical Engineering* 6, no. 07 (2021): 11–27.
16. Yesankar, Prajyot R., Prajwal Bandre, Prateek Verma, Amit G. Gudadhe, Palash Gourshettiwar, and Chetan Puri. "A review on the role of IoT in smart agriculture with reference to efficiency, sustainability and precision farming." In *2024 5th International Conference on Electronics and Sustainable Communication Systems (ICESC)*, pp. 533–537. IEEE, 2024.
17. Jaiswal, Shiva Pujan, Vikas Singh Bhadoria, Amit Agrawal, and Hemant Ahuja. "Internet of Things (IoT) for smart agriculture and farming in developing nations." *International Journal of Scientific & Technology Research* 8, no. 12 (2019): 1049–1056.
18. Hasan, Md Mahadi, Muhammad Usama Islam, and Muhammad Jafar Sadeq. "Towards technological adaptation of advanced farming through AI, IoT, and Robotics: A Comprehensive overview." *arXiv preprint arXiv:2202.10459* (2022).
19. Garg, Satvik, Pradyumn Pundir, Himanshu Jindal, Hemraj Saini, and Somya Garg. "Towards a multimodal system for precision agriculture using IoT and machine learning." In *2021 12th International Conference on Computing Communication and Networking Technologies (ICCCNT)*, pp. 1–7. IEEE, 2021.
20. Rabha, Garima, Kundan Kumar, Dhruv Kumar, and Diwakar Kumar. "A comprehensive review of integrating AI and IoT in farm machinery: Advancements, applications, and sustainability." *Environmental Pollution* 17 (2023): 18.
21. Mana, A. A., A. Allouhi, A. Hamrani, S. Rahman, I. el Jamaoui, and K. Jayachandran. "Sustainable AI-based production agriculture: Exploring AI applications and implications in agricultural practices." *Smart Agricultural Technology* 7 (2024): 100416.
22. Maheshwari, Rohit, Mohnish Vidyarthi, and Parth Vidyarthi. "Significant role of IoT in agriculture for smart farming." In *IoT and AI Technologies for Sustainable Living, A Practical Handbook,* edited by Abid Hussain, Garima Tyagi, and Sheng-Lung Peng. pp. 43–55. CRC Press, 2022.
23. Ryan, Mark. "The social and ethical impacts of artificial intelligence in agriculture: Mapping the agricultural AI literature." *AI & SOCIETY* 38, no. 6 (2023): 2473–2485.
24. Adewusi, Adebunmi Okechukwu, Onyeka Franca Asuzu, Temidayo Olorunsogo, Chinwe Iwuanyanwu, Ejuma Adaga, and Donald Obinna Daraojimba. "AI in precision agriculture: A review of technologies for sustainable farming practices." *World Journal of Advanced Research and Reviews* 21, no. 1 (2024): 2276–2285.
25. Hingle, Ashish, and Aditya Johri. "Recognizing principles of AI ethics through a role-play case study on agriculture." In *2023 ASEE Annual Conference & Exposition*. ASEE Conferences, 2023.
26. Sparrow, Robert, and Mark Howard. "Robots in agriculture: Prospects, impacts, ethics, and policy." *Precision Agriculture* 22 (2021): 818–833.
27. Jawale, Deepali, and Sandeep Malik. "Smart farming technology with AI & block chain: A review." In *Smart Computing Techniques and Applications: Proceedings of the Fourth International Conference on Smart Computing and Informatics, Volume 2*, pp. 757–763. Springer Singapore, 2021.
28. Chowdhury, Hribhu, Debo Brata Paul Argha, and Md Ashik Ahmed. "Artificial intelligence in sustainable vertical farming." *arXiv preprint arXiv:2312.00030* (2023).
29. Pranto, Tahmid Hasan, Abdulla All Noman, Atik Mahmud, and AKM Bahalul Haque. "Blockchain and smart contract for IoT enabled smart agriculture." *PeerJ Computer Science* 7 (2021): e407.
30. Puri, Bhuvan, and Rameshwar Cambow. "A review on the role of IoT, AI, and blockchain in agriculture & crop diseases detection using a text mining approach." *Ingeniería Solidaria* 19, no. 3 (2023): 1–30.

31. Qazi, Sameer, Bilal A. Khawaja, and Qazi Umar Farooq. "IoT-equipped and AI-enabled next generation smart agriculture: A critical review, current challenges and future trends." *IEEE Access* 10 (2022): 21219–21235.
32. Liu, Ye, Xiaoyuan Ma, Lei Shu, Gerhard Petrus Hancke, and Adnan M. Abu-Mahfouz. "From industry 4.0 to agriculture 4.0: Current status, enabling technologies, and research challenges." *IEEE Transactions on Industrial Informatics* 17, no. 6 (2020): 4322–4334.
33. Jha, Kirtan, Aalap Doshi, Poojan Patel, and Manan Shah. "A comprehensive review on automation in agriculture using artificial intelligence." *Artificial Intelligence in Agriculture* 2 (2019): 1–12.
34. Balafoutis, Athanasios, Bert Beck, Spyros Fountas, Jurgen Vangeyte, Tamme Van der Wal, Iria Soto, Manuel Gómez-Barbero, Andrew Barnes, and Vera Eory. "Precision agriculture technologies positively contributing to GHG emissions mitigation, farm productivity and economics." *Sustainability* 9, no. 8 (2017): 1339.
35. Kamilaris, Andreas, and Francesc X. Prenafeta-Boldú. "Deep learning in agriculture: A survey." *Computers and Electronics in Agriculture* 147 (2018): 70–90.
36. Wang, Jun, Yanlong Wang, Guang Li, and Zhengyuan Qi. "Integration of remote sensing and machine learning for precision agriculture: A comprehensive perspective on applications." *Agronomy* 14, no. 9 (2024): 1975.
37. Mishra, Harshit, and Divyanshi Mishra. "Sustainable smart agriculture to ensure zero hunger." In Saravanan Krishnan, A. Jose Anand, and Raghvendra Kumar (eds.). *Sustainable Development Goals*, pp. 16–37. CRC Press, 2024.
38. Pattanayak, Sandhya, Arpan Kumar Basak, Satyaki Ghosh, Anilesh Dey, Saradindu Panda, Surajit Bari, Kaushik Sarkar et al. "AI and IoT based smart irrigation system." *Turkish Online Journal of Qualitative Inquiry* 12, no. 6 (2021).
39. Rani, Ruchi, Jayakrushna Sahoo, Sivaiah Bellamkonda, Sumit Kumar, and Sanjeev Kumar Pippal. "Role of artificial intelligence in agriculture: An analysis and advancements with focus on plant diseases." *IEEE Access* 11 (2023): 137999–138019.

9 A Scalable Deep Learning Framework for Cotton Crop Disease Diagnosis in Resource-Limited Environments

Shashank M P, Shalini M S, Shashidhar R, and Roopashree S

9.1 INTRODUCTION

In many countries, including India, one of the most significant occupations is agriculture. In the agricultural industry, crop infection detection has become increasingly crucial from a technical and financial standpoint. A system that can automatically identify diseases is necessary because it can revolutionize the monitoring of vast agricultural fields and enable the prompt treatment of plant leaves once an illness is detected.

Cotton is a raw material of the textile industry, and it is given priority as compared to other cash crops. It is the livelihood of around 60 million people and is also a major agribusiness, with billions of farmers across countries earning a profit out of the agricultural product. Additionally, this work provides valuable information about automatically identifying diseases on leaves. Field crops present certain challenges, though, such as determining the different pests and diseases that affect the crops and assessing plant nutritional deficiencies.

Even though each issue is significant, identifying the pests to take appropriate measures in order to reduce loss is the most crucial of the aforementioned issues. Farmers become aware of a particular condition and take appropriate action to control it whenever it occurs. However, if agriculturists lacks the necessary information, they may choose ineffective control methods, like pesticides, wasting time and money. It also causes a major problem with crops. The diagnostic should possess excellent observational skills and be a competent investigator. Therefore, there is a need to keep an open mind until then when the relevant information concerning the issue is obtained. It should also consider the possibility of many causative factors. Also, the controls solely rely on proper diagnosis of the diseases and the disease-causing organisms. Hence, the diagnosis of diseases can be accorded the utmost priority in the education of a plant pathologist. Moreover, the control measures shall stage deterioration to the plant in addition to losing time and money in case the disease and the agent causing it are not well identified.

Farm owners can seek assistance from agricultural experts, who would advise the farmers on useful hints on how to identify diseases and enhance production. Nonetheless, some of the anticipated common problems for the farmers include the following: failure of the expert that the farmer intends to consult to offer pertinent knowledge and information, failing to access help because of long distances, or even the chance of lack of availability of the agriculture expert when he/she visits.

These days, artificial intelligence (AI) analyzers have improved to effectively classify and detect leaf diseases automatically through image processing. These days, researchers prefer to use deep

DOI: 10.1201/9781003613510-9

learning in neural networks. The most popular option is deep learning because of its crucial ability to automatically extract features from the provided images. In order to automatically extract features, they must be trained. A preferred model is the multi-layered Convolutional Neural Network (CNN). These networks have been used to solve image classification problems in recent years because of their efficiency and simplicity of use. With 70% of the Indian textile market, it holds the top spot despite fierce competition from synthetic fibers. Therefore, the country's textile industry and GDP will be impacted by the loss of cotton crops due to unprecedented crop diseases.

Thus, early detection of cotton leaf or plant disease can help farmers avoid massive debt and support India's textile sector. Problem statement of the proposed work: Formulate a custom model of CNN model that helps in the detection and classification of diseases, and a model that could be used on an Android application to classify plant or cotton leaf diseases in real time.

The proposed work's primary contributions are

- To compile fresh and diseased cotton plant photos from public sources into a standard dataset.
- To add more images to the dataset by using image preprocessing methods like image augmentation, to improve accuracy by applying image scaling, rotation, and resizing as well.
- To import every package and library needed to put the CNN into practice.
- To create a unique deep-layer CNN architecture with the right settings to achieve high accuracy.
- To compare the outcomes with a custom CNN model and analyze the results using transfer learning models like Resnet50 and InceptionV3.
- To incorporate the model into an Android application for disease detection in real time.

9.2 LITERATURE SURVEY

The authors proposed a model for cotton leaf disease detection and recognition with an accuracy of 96%. A main demerit of the work is it takes a lot of time for preprocessing [1]. A deep convolution network is used for the implementation with an accuracy of 96% and in this work the authors used very little data for processing [2, 3]. Kiani and Tofik described the plant disease detection and recognition for modern agriculture, and the authors verified the novel method based on computer vision to detect the strawberry leaves [4]. Rauf et al. created the citrus plant database, then applied the different machine learning algorithms for analysis. The authors created a healthy and different disease database for this work [5]. Saradhamba et al. worked on identifying infected leaf and stem, Portion the pretentious part, discovering the form of diseased area, controlling the color of diseased area, and inspiration of size and form of greenery [6]. K-means clustering is used for partitioning the leaf image. Classification was done based on a neural network using a back propagation algorithm [7]. The SVM algorithm was used, achieving an 82% overall system accuracy. Three feature vectors were retrieved, like color histogram, morphology, and CCV feature vectors, each [8]. The author used Combined MobileNet+VGG16 characteristics to address the issue of class imbalance, and the ensemble increases the robustness [9]. The highest performance was achieved using ResNet152; the advantage was due to the use of an image dataset that was collected in the field, and also a broader model comparison was used [10]. The paper used a DCNN, which improved the accuracy with pooling/layer tuning. The dataset images of 2,293 were used. Formal TLA + model for correction is used with an accuracy of 97.88% increasing the reliability [11]. A Deep CNN on ResNet-15V2 transfer learning was employed, which helped in robust classification, achieving a higher precision in identifying multiple disease types [12]. A combinational classifier, VGG 16, DendeNet201, ResNet50, Transfer learning with GoogleNet, is adapted in this research; for transfer learning – ResNet 152V2 data normalization and augmentation was used, reaching about 92% accuracy [13]. The MobileNet subset + augmentation with a low parameter count is suitable for deployment on a resource-limited device. The accuracy of 98.42 was reached; the downside was that the use of a lesser disease type had been tested, and for validation, a larger field dataset was required [14]. This survey used 121 CNN-based plant models with a broader taxonomy; the advancement identifies trends and gaps for future study [15]. Thakur et al. used a hybrid CNN with a lightweight architecture, also an advantage of IoT, IoT-friendly, good generalization across multiple crops few downsides are cotton plant-specific dataset is missing [16]. Mohanty et al. used CNN and trained on the publicly available dataset, achieving a 99.35% accurate result on the test dataset with application to mobile interface [17]. The research uses spectral, thermal, and hyperspectral images with a multimodal deep network, which uses a multi-sensor functionality that enables early detection beforehand [18]. Azfar et al. used a pre-trained model, VGG16 and CNN (custom), with Inception V3 on leaves and bolls using IoT with the leverage of real-time IoT enabled with a wider class of data [19]. The classification used transfer learning, VGG19, Inception V3, and ResNet on the standard cotton dataset, where ResNet 152V2 achieved an accuracy of 98.63% with a few downsides, like a bulky model and having an older dataset [20]. A Hybrid ML and CNN model was used with the feature selection, obtaining an optimized accuracy, but it is still in an experimental stage, and the details about deployment are unclear [21]. The research used different activation functions with a CNN algorithm, which fine-tuned the performance, and also improved convergence and stability, but there was no real dataset deployment, thus giving a narrow focus [22]. The authors used CNN to compare training on synthetic versus original images, showing a better efficiency of the synthetic augmentation, having a limited disease class [23]. A few different types of algorithms, like CNN, hybrid model, GAN, and TL for plant disease vision systems, were employed, which highlight trends and address all the research gaps. However, they lack an experimental framework, and the research is not specific to cotton leaf disease [24].

9.3 METHODOLOGY

9.3.1 Dataset

Pictures of cotton plants were gathered from a variety of public sources. The dataset's images are taken in a variety of lighting and intensity settings. In our dataset, we have selected four image classes: diseased cotton leaves, diseased cotton plants, fresh cotton leaves, and fresh cotton plants.

The training dataset consists of 2,000 images in total. There are 100 images for validation and 500 images for testing. The Cotton Leaf Disease Detection Dataset is one of the resources that can help those who are researching, practicing, or concerned with agricultural activities to better manage the issues with cotton leaf diseases since it provides researchers and stakeholders with handy information and tools that can be applied to this situation. The dataset can facilitate early identification of diseases in cotton leaves during the formation of appropriate remedial measures, and be a useful technology to the farming community, so that appropriate measures can be implemented in time and crop management procedures given the optimum. Furthermore, it enhances the development of tools and methods in the detection of diseases, which creates new perspectives in agricultural research. The identified dataset is the collection of carefully selected photographs at different points of the cotton leaf diseases taken in the "National Cotton Research Institute field in Gazipur." Images that are taken with a phone have a wide variety of representations of disease manifestations on various levels. Though field surveys between October 2023 and January 2024 had to deal with the limitation of changing lighting conditions, the work was based on the guidance of domain experts to guarantee the quality collection of images. It consists of 2,137 images of seven classes referring to different states of cotton leaves, such as bacterial blighted, curl virus-infected, and healthy leaves. Within these classes, the issues that cotton plants may face can be put forward as diseases, pests, and environmental stress. There is also the meticulous preparation and augmentation of the dataset, involving cleaning the data, labeling, and enhancement through approaches like flipping and brightening. Therefore, it contains a total of 2,137 original plus 7,000 augmented images that can yield even better deep learning models to enable the accurate classification and diagnosis of the diseases.

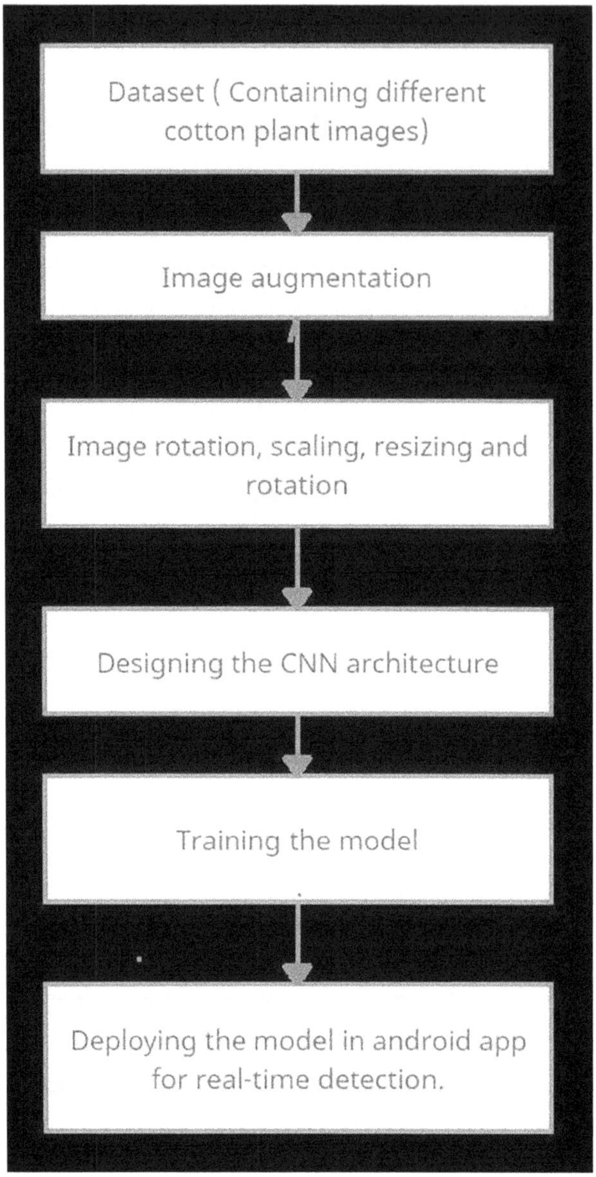

FIGURE 9.1 Proposed system block.

9.3.2 Image Augmentation

Neural networks involved in deep learning tend to get better as more data is used. Data augmentation is the process of utilizing preexisting training data in order to create new instances artificially. This is done by generating fresh and unique training examples using domain-specific methods on training data examples.

Transforming original versions of the images in the training set to different but same-class creations thereof is called image data augmentation, and is the most well-known type of data augmentation. Shifts, flips, zooms, and many other image manipulation techniques are examples of transforms. It is used to add fresh, credible examples to the training dataset. This indicates that the

FIGURE 9.2 Image of a diseased cotton leaf.

model is likely to see different versions of the training set image. An example would be a picture of a cat; this would be reasonably upside down since it may be shot from the left or the right.

Data augmentation used for the training dataset should be considered, taking into account the training dataset's context as well as the problem of the domain knowledge. Additionally, testing data augmentation techniques separately and together can be helpful to determine whether they improve model performance in a quantifiable way. This can be performed using a small prototype data set, model, and running the training.

Image data augmentation is normally carried, as in Figure 9.4, out only on the training dataset. The validation or test data is not applied to augmentation. It is not like the activities of data preparation used to support the work, such as pixel scaling and image resizing, which have to occur harmoniously in all the data sets that interact with the model.

9.3.3 Image Rotation, Scaling, and Resizing

Rotation of an image for an angle θ is achieved by the transformation matrix of the form.

$$M = \begin{bmatrix} Cos\,\theta & -Sin\,\theta \\ Sin\,\theta & Cos\,\theta \end{bmatrix} \tag{9.1}$$

Keras offers scaled rotation around the point of your choice so that you are allowed to rotate anywhere you wish. The transformation matrix in modified form is given by

FIGURE 9.3 Image of a healthy cotton leaf.

$$\begin{matrix} \alpha\beta(1-\alpha).center.x-\beta.center.y \\ -\beta\ \alpha\beta.center.x+(1-\alpha).center.y \end{matrix} \tag{9.2}$$

9.3.3.1 Design of CNN Architecture

Neural networks called CNNs can detect the various aspects of parameters within an image by providing the weights or biases to them. The architecture of the Visual Cortex was the prototype of the architecture of the ConvNet, that is, similar to the way neurons connect in the human brain. Individual neurons only respond to the stimulus at a relatively restricted part of the visual field called the Receptive Field. There is an overlap of the fields that cover the entire visual area.

$$(f*g)(\mathrm{t}) = \int_{-\infty}^{\infty} f(\tau)g(t-\tau)d\tau \tag{9.3}$$

The task of ConvNet is to downsize images to a level that is easier to work on, yet preserving characteristics that can be deemed as crucial in obtaining a decent prediction. When architecture is developed, good learning is incorporated for a larger dataset.

$$n_{out} = \left[\frac{n_{in}+2p-k}{s}\right]+1 \tag{9.4}$$

FIGURE 9.4 Image augmentation.

where n_{in} = number of input features; n_{out} = number of output features; k = Conv kernel size.

Extraction of the input image's high-level features, like edges, is the aim of the convolution operation. Convolutional networks do not have to be restricted to a single Conv Layer. One would expect the first ConvLayer to capture low-level features, as shown in Figure 9.5. Architecture adapts to high-level features as well, with an extra layer, providing us with a network that understands the images in the dataset holistically, just as we would.

9.3.3.2 Fully Connected Layers

The final few layers of the network are fully connected layers. Output of the final pooling or convolutional layer flows into the fully connected layer that is flattened before being fed.

$$Z = W^T.X + b \tag{9.5}$$

$$Z = \begin{bmatrix} w_{11}\ w_{21}\ w_{31}\ w_{41} \\ w_{12}\ w_{22}\ w_{32}\ w_{42} \end{bmatrix} \begin{bmatrix} X_1 \\ X_2 \\ X_3 \\ X_4 \end{bmatrix} + \begin{bmatrix} b_1 \\ b_2 \end{bmatrix} = \begin{bmatrix} w_{11}\ w_{21}\ w_{31}\ w_{41} \\ w_{12}\ w_{22}\ w_{32}\ w_{42} \end{bmatrix} \begin{bmatrix} X_1 \\ X_2 \\ X_3 \\ X_4 \end{bmatrix} + \begin{bmatrix} b_1 \\ b_2 \end{bmatrix}$$

$$Z_{2\times2} = \begin{bmatrix} w_{11}\ X_1 + w_{21}\ X_2 + w_{31} X_3 + w_{41} X_4 \\ w_{12}\ X_1 + w_{22}\ X_2 + w_{32} X_3 + w_{42} X_4 \end{bmatrix} \tag{9.6}$$

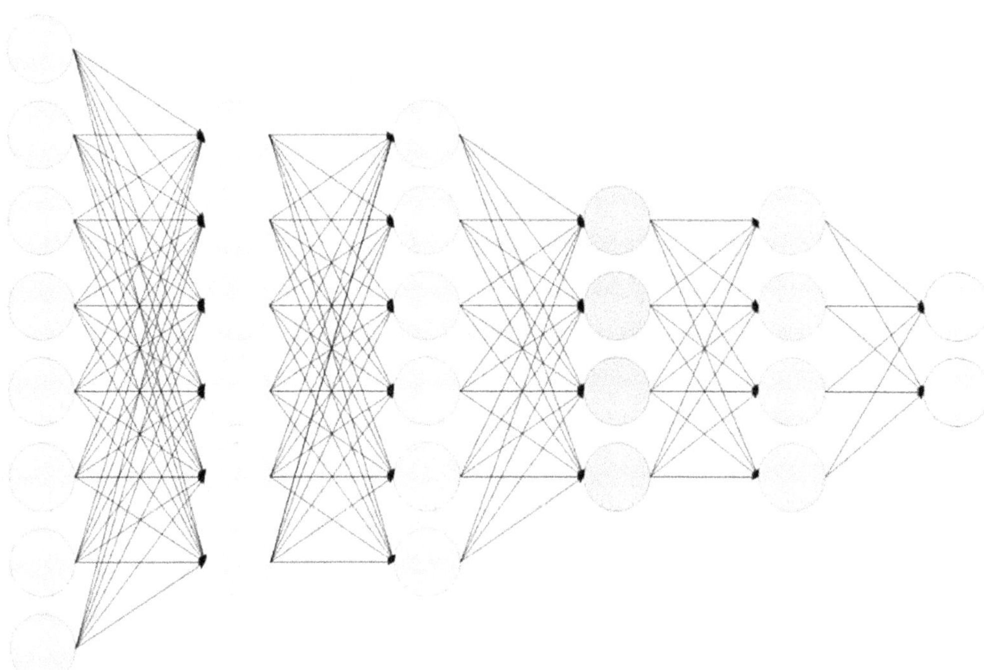

FIGURE 9.5 Fully connected layer.

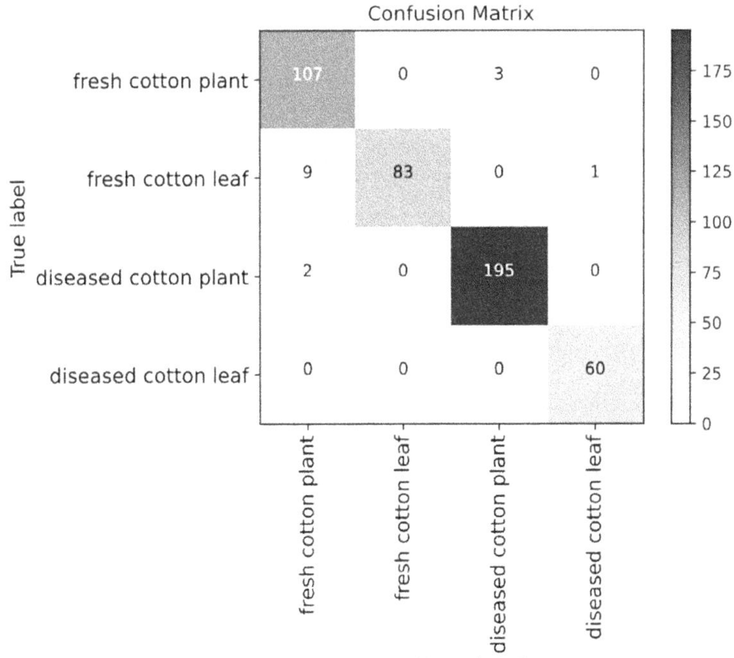

FIGURE 9.6 Confusion matrix.

This confusion matrix demonstrates the fact that the classification model works extremely well, and most of the predictions are correct among all four classes (fresh/diseased cotton plant/leaf). There are a few misclassifications, meaning that it is very accurate, as shown in Figure 9.6.

Accuracy in %	Precision	Recall	F1 Score
95.89	0.94	0.99	0.97
99.91	0.99	0.99	0.99

The model is highly accurate, with an accuracy score close to 99.91%. Other scores are close to 0.99 as well. This is substantiated by the confusion matrix, which indicates that the majority of samples are properly classified, with a very small number of mistakes in all types of cotton plants and leaves, as shown in Table 9.1.

9.3.4 Algorithm

Step 1: Import Tensorflow/Keras libraries and packages
Step 2: Partition the dataset into different classes and upload the dataset
Step 3: Apply image preprocessing techniques like image augmentation, image rotation, scaling, and resizing.
Step 4: Develop the CNN architecture with an adequate number of layers in terms of filter, dense, and pooling layers.
Step 5: Train the model within an adequate number of epochs, batch size, and learning rate
Step 6: Train the CNN model using different test images and scale up the model with different parametric values
Step 7: Deploy the model into the Android app.

TABLE 9.1
Classification Table

Model: "sequential"

Layer (type)	Output Shape	Param #
conv2d (Conv2D)	(None, 148, 148, 32)	896
max pooling2d (MaxPooling2D)	(None, 74, 74, 32)	0
conv2d_1 (Conv2D)	(None, 72, 72, 64)	18496
max_pooling2d_1 (MaxPooling2	(None, 36, 36, 64)	0
conv2d 2 (Conv2D)	(None, 34, 34, 128)	73856
max_pooling2d_2 (MaxPooling2	(None, 17, 17, 128)	0
conv2d 3 (Conv2D)	(None, 15, 15, 256)	295168
max_pooling2d_3 (MaxPooling2	(None, 7, 7, 256)	0
dropout (Dropout)	(None, 7, 7, 256)	0
flatten (Flatten)	(None, 12544)	0
dense (Dense)	(None, 128)	1605760
dropout_l (Dropout)	(None, 128)	0
dense _1 (Dense)	(None, 256)	33024
dropout_2 (Dropout)	(None, 256)	0

9.4 CONCLUSION AND FUTURE SCOPE

This chapter has effectively depicted how transfer learning-based pipeline is used for breast cancer detection by using a model ResNet-50 along with several preprocessing strategies (i.e., thresholding and pectoral muscle masking) to enhance image clearness and classification precision. It has shown a good performance with the training, validation, and test accuracy of over 92%, and sensitivity in detecting malignant cases that are pivotal in early diagnosis. These findings suggest the promising nature of CNN-based methods to perform accurate and explainable medical imaging analysis. Going forward, it is possible to expand the data range of this work with more samples, including multi-institutional and multiracial ones, and fine-tuning hyperparameters, and thoroughly investigate deep architectures to achieve even better performance. Robustness of diagnosis can be increased by the integration of multimodal data like ultrasound, MRI, genetic data, etc. Also, mobile or cloud-based platform lightweight deployment and automated preprocessing pipelines can empower real-time, scalable, and accessible clinical use case. In summary, the suggested model establishes a basic framework of feasible and AI-supported diagnostic systems that have the potential of supporting radiologists in an early and precise breast cancer detection, which will benefit patients overall.

REFERENCES

[1] K. Prashar, R. Talwar, and C. Kant, "CNN Based on Overlapping Pooling Method and Multi-layered Learning with SVM & KNN for American Cotton Leaf Disease Recognition," *2019 International Conference on Automation, Computational and Technology Management (ICACTM)*, London, UK, 2019, pp. 330–333, doi: 10.1109/ICACTM.2019.8776730.

[2] S. S. Hari, M. Sivakumar, P. Renuga, S. Karthikeyan, and S. Suriya, "Detection of Plant Disease by Leaf Image Using Convolutional Neural Network," *2019 International Conference on Vision Towards Emerging Trends in Communication and Networking (ViTECoN),* Vellore, India, 2019, pp. 1–5, doi: 10.1109/ViTECoN.2019.8899748.

[3] A. Jenifa, R. Ramalakshmi and V. Ramachandran, "Cotton Leaf Disease Classification using Deep Convolution Neural Network for Sustainable Cotton Production," *2019 IEEE International Conference on Clean Energy and Energy Efficient Electronics Circuit for Sustainable Development (INCCES),* Krishnankoil, India, 2019, pp. 1–3, doi: 10.1109/INCCES47820.2019.9167715.

[4] W. Chen, C. He, C. Ji et al. "An Improved K-means Algorithm for Underwater Image Background Segmentation," *Multimed Tools Appl., vol.* 80, 21059–21083, 2021. https://doi.org/10.1007/s11042-021-10693-7.
[5] E. Kiani and T. Mamedov, "Identification of Plant Disease Infection Using Soft-Computing: Application to Modern Botany," *Procedia Comput. Sci., vol.* 120, 893–900, 2017. https://doi.org/10.1016/j.procs.2017.11.323.
[6] H. T. Rauf, B. ALi Saleem, M. I. Lali, and M. Attique Khan, "A Citrus Fruits and Leaves Dataset for Detection and Classification of Citrus Diseases through Machine Learning," *Data in Brief.* 26. 104340. 10.1016/j.dib.2019.104340.
[7] D. Saradhambal and L. R. Rajesh, "Plant Disease Detection and Its Solution Using Image Classification," *Int. J. Pure Appl. Math.,* 2018, vol. 9, pp. 118–124.
[8] G. K. Sandhu and R. Kaur, "Plant Disease Detection Techniques: A Review," *2019 International Conference on Automation, Computational and Technology Management (ICACTM)*, London, UK, 2019, pp. 34–38, doi: 10.1109/ICACTM.2019.8776827.
[9] A. Aslam, S. M. Usman, M. Zubair, and I. Hussain, "Multi-Convolutional Neural Networks for Cotton Disease Detection Using Synergistic Deep Learning Paradigm," *PLOS One*, vol. 20, no. 5, e0324293, May 2025. https://doi.org/10.1371/journal.pone.0324293.
[10] H. M. Faisal, M. Aqib, S. U. Rehman et al. "Detection of Cotton Crops Diseases Using Customized Deep Learning Model," *Sci Rep., vol.* 15, 10766, 2025. https://doi.org/10.1038/s41598-025-94636-4.
[11] A. Rehman, N. Akhtar, and O. H. Alhazmi, "Monitoring and Predicting Cotton Leaf Diseases Using Deep Learning Approaches and Mathematical Models," *Sci Rep.*, vol. 15, 22570, 2025. pp. 98–105 https://doi.org/10.1038/s41598-025-06985-9.
[12] P. Chaudhari, R. V. Patil, and P. N. Mahalle, "An Intelligent Approach for Cotton Plant Disease Detection Using CNNs: ResNet-152V2," *J. Electr. Syst. Res.*, vol. 20, no. 1s, Mar 2024. pp. 112–121.
[13] P. Chaudhari, R. V. Patil, and P. N. Mahalle "Deep Learning-Based Cotton Plant Disease Detection Using CNNs," *JISEM J.*, vol. 10, no. 22s, 2025. https://doi.org/10.52783/jisem.v10i22s.3618.
[14] A. K. Patra and T. Gajurel, "Improved Cotton Leaf Disease Classification Using Parameter-Efficient Deep Learning Framework," arXiv preprint Dec 2024. https://doi.org/10.48550/arXiv.2412.17587.
[15] A. Abade, P. Ferreira, and F. Vidal, "Plant Diseases Recognition on Images Using Convolutional Neural Networks: A Systematic Review," 2020. doi: 10.48550/arXiv.2009.04365.
[16] P. S. Thakur et al., "Explainable Vision Transformer Enabled CNN for Plant Disease Identification: PlantXViT," arXiv, article number 1419, Jul 2022. https://doi.org/10.48550/arXiv.2207.07919.
[17] S. P. Mohanty and P. Hughes David, Salathé Marcel "Using Deep Learning for Image-Based Plant Disease Detection," *Front. Plant Sci.,* vol. 7, 2016 www.frontiersin.org/journals/plant-science/articles/10.3389/fpls.2016.01419.
[18] A. R. Prodeep, A. S. M. Morshedul Hoque, Md. S. Kabir, Md. S. Rahman, and M. Firoz Mridha, "Plant Disease Identification from Leaf Images Using Deep CNN's EfficientNet," *Journal of Artificial Intelligence and Capsule Networks*, March 2025, vol. 7, no. 1, pp. 32–50, 2022. https://doi.org/10.1109/DASA54658.2022.9765063.
[19] S. Azfar et al., "Automated System for Detecting, Identifying, and Preventing Cotton Leaf and Boll Diseases Using Deep Learning," *Computers and Electronics in Agriculture.* vol. 155, pp. 220–236. Mar 2025.
[20] S. Perumal, N. Mohan, P. Kavya, and P. Teja, "Leaf Disease Identification: Enhanced Cotton Leaf Disease Identification Using Deep CNN Models," *International Conference on Intelligent Systems, Smart and Green Technologies (ICISSGT),* 22–26, 2021. https://doi.org/10.1109/ICISSGT52025.2021.00016.
[21] X. Sun, G. Li, P. Qu, X. Xie, X. Pan, and W. Zhang, "Research on Plant Disease Identification Based on CNN," *Cognitive Robotics.,* vol. 2, 155–163, 2022. https://doi.org/10.1016/j.cogr.2022.07.001.
[22] K. Joshitha, M. Subramanian, L. Shakthi, and L. Prasanth, "Design and Comparison of Activation Functions for the Classification of Cotton Leaf Diseases," *International Conference on Power, Energy, Control and Transmission Systems*, 1–4, 2024. https://doi.org/10.1109/ICPECTS62210.2024.10780182.
[23] P. S. Bolisetty, N. A. Lakkshmi Yogesh , S. D. Kosuri, P. Sravan, S. V. Sai Karthik, and C. B. Rajesh, "Comparative Analysis of CNN Models' Accuracy Using Synthetic and Original Datasets

for Cotton Leaf Classification," *2024 4th International Conference on Artificial Intelligence and Signal Processing (AISP), VIJAYAWADA, India*, 2024, pp. 1–5, https://doi.org/10.1109/AISP61711.2024.10870706.

[24] R. S. Joseph, C. A. Nisha, and K. K. Shandry., "Plant Disease Detection: A Survey on Recent Deep Learning Technologies," *2022 Second International Conference on Next Generation Intelligent Systems (ICNGIS)*, Kottayam, India, 2022, pp. 1–6, doi: 10.1109/ICNGIS54955.2022.10079766.

10 Next-Generation Agriculture

Balakrishnan D, Aarthy C, Renuga Devi B,
Vasuki R, and Narmada C

10.1 INTRODUCTION

The global agrarian sector is facing a period of significant metamorphosis, driven by rapid-fire population growth, climate change, and resource constraints. With the world population projected to outreach 9 billion by 2050, the dun for food is anticipated to rise exponentially, putting immense stress on agrarian systems to enhance productivity while maintaining sustainability. Still, traditional husbandry styles alone are inadequate to meet this growing demand, as they're decreasingly challenged by climate variability, soil declination, water failure, and labor deaths. These complex challenges bear the relinquishment of coming-generation agrarian practices that influence slice-edge technologies, and sustainable approaches to ameliorate effectiveness, and adaptability.

In recent times, technological advancements similar to perfection husbandry, perpendicular husbandry, and the integration of artificial intelligence (AI), the Internet of Things (IoT), and biotechnology hold revolutionized bread production networks. Precision husbandry, for case, utilizes data-driven perceptivity, and IoT-enabled bias to optimize resource operation, ensuring that water, diseases, and fungicides are applied efficiently to maximize crop yields while minimizing waste. This approach enables growers to make informed opinions grounded on real-time data about soil health, rainfall conditions, and crop performance, eventually leading to advanced productivity, and reduced environmental impact, perpendicular husbandry, and controlled-terrain husbandry Controlled Environment Agriculture (CEA) present innovative systems for addressing land and resource constraints.

These styles involve growing crops in piled layers or controlled inner surroundings, allowing for time-round food products independent of external climate conditions. By exercising hydroponic or aeroponic systems, perpendicular husbandry significantly reduces water operation compared to traditional soil-grounded husbandry, making it an ideal result for civic food products. Also, civic husbandry enterprises are gaining traction as metropolises seek to enhance local food security and reduce reliance on long-distance food transportation, contributing to carbon emigration. Artificial intelligence and machine literacy are playing a decreasingly vital part in ultramodern husbandry, particularly in the realm of remote sensing, and automated decision tree. AI-powered tools dissect satellite, and drone imagery to describe crop health issues, pest infestations, and nutrient scarcities before they escalate, enabling growers to take visionary measures to cover their yields. These technologies also contribute to prophetic analytics, allowing for accurate yield soothsaying, and optimized resource allocation. Integrating AI in husbandry operations enhances perfection, effectiveness, and sustainability, eventually leading to more flexible agrarian practices.

The IoT further augments smart husbandry by deploying connected detectors and automated systems. IoT bias similar to soil humidity detectors, climate observers, and independent irrigation systems

DOI: 10.1201/9781003613510-10

enables growers to ever cover, and manage their fields with unknown perfection. By automating critical ranch operations, IoT-driven husbandry minimizes mortal labor conditions, reduces functional costs, and ensures optimal growing conditions for crops. These advancements are necessary in mollifying the impact of climate change on food products and promoting resource-effective husbandry methods.

Biotechnology and synthetic biology have also surfaced as game changers in coming-generation husbandry. Genetically modified organisms and gene-editing technologies have allowed the progress of harvests with enriched resistance to pestilences conditions, and extreme rainfall conditions. These inventions contribute to food security by adding crop yields and perfecting the nutritive value of agrarian products. Also, synthetic biology holds a pledge to develop sustainable druthers to chemical diseases, and fungicides, further reducing the ecological footmark of ultramodern husbandry practices. Despite the promising eventuality of these technological advancements, challenges remain in icing-wide relinquishment and availability.

Digital inequality poses a significant hedge, particularly in developing regions where access to high-speed internet, advanced husbandry outfits, and fiscal coffers is limited. Smallholder growers, who constitute a significant portion of the global agrarian pool, frequently struggle to make the original investment needed for espousing smart husbandry technologies. Likewise, nonsupervisory fabrics and ethical considerations girding biotechnology, and data sequestration must be precisely addressed to ensure the responsible perpetration of these inventions.

The future of husbandry lies in striking a balance between technological progress and sustainability. Governments, exploration institutions, and private sector stakeholders must unite to produce programs, and structures that support the integration of coming-generation agrarian systems. Investments in digital structure, planter education, and sustainable husbandry enterprise will be pivotal in bridging the technological gap and making these advancements accessible to all growers, regardless of geographic or profitable limitations.

Coming-generation husbandry represents a transformative shift toward data-driven, technology-powered, and environmentally sustainable food product systems. By employing the eventuality of AI, IoT, biotechnology, and perfect husbandry, the agrarian sector can overcome challenges and ensure food security for future generations. Still, the success of these inventions will depend on indifferent access, robust policy support, and continued exploration to enhance effectiveness, scalability, and sustainability. As the world moves toward a further connected and digitally advanced future, embracing coming-generation husbandry will be essential in erecting a flexible, effective, and sustainable global food system.

The remainder of this chapter is organized as follows: Section 10.2 provides a complete consideration of subsisting literature and related works that inform the development of next-generation agricultural systems. Section 10.3 delves into the integration of fragile sensing technologies and AI in precision husbandry, emphasizing their transformative role in authentic monitoring. Section 10.4 explores the applications of the IoT in farming, highlighting its contributions to operational efficiency and sustainability. Section 10.5 discusses inventions in perpendicular husbandry, controlled-environment agriculture, and urban farming as sustainable solutions to land scarcity and climate variability. Section 10.6 examines sustainable pest management approaches through the deployment of biopesticides and bioinoculants. Section 10.7 outlines the prevailing challenges in technology adoption, digital inequality, and the necessity for inclusive policy frameworks. Section 10.8 underscores the critical need for next-generation agriculture in addressing food security, climate change, and resource constraints. Finally, Section 10.9 presents the conclusion, recapitulating crucial perceptivity and suggesting future exploration directions.

10.2 RELATED WORKS

Abdi and Barwant (2025) explore the role of microbial and algal biostimulants in fostering sustainable agriculture, emphasizing their potential to enhance crop growth, improve soil health, and increase stress adaptability. Their analysis highlights how biostimulants can reduce dependency on chemical inputs, thereby promoting eco-friendly, and resilient farming practices. Bampasidou et al. (2024) emphasize that integrating digital technologies into agriculture is crucial for bridging the digital divide, particularly in rural areas. They highlight the critical role of higher education in cultivating the next generation of digital agriculture professionals, equipping them with essential skills to effectively harness and apply emerging technologies for precision farming, smart decision-making, and sustainable agricultural development.

Gebregziabher et al. (2024) probe genomic regions, and seeker genes linked to carotenoid pile up in soybeans applying nearing origination sequencing-predicated core segregate analysis. Their study enhances the understanding of inheritable factors impacting carotenoid content, abetting soybean bio-fortification, and crop enhancement sweats. MondějarLópez et al. (2024) review the use of natural polymer-grounded nanoparticles for recapitulating eco-friendly agrochemicals, enhancing crop protection, and sustainability. They punctuate benefits like bettered effectiveness, reduced environmental impact, and controlled release in husbandry.

Varghese et al. (2024*)* present an AI-grounded eco-friendly quad-copter model designed for sustainable husbandry. Their study explores how AI-driven drones can optimize crop monitoring, pest control, and resource operation, enhancing effectiveness while reducing environmental impact. Garg et al. (2024) claw coming- birth factory biostimulants, and genome sequencing arrangements for a tenable frugality. They punctuate the part of biostimulants in enhancing crop growth, and stress forbearance while agitating genomic approaches to ameliorate factory adaptability and productivity.

Mahobia et al. (2024*)* explore the part of proteomics in advancing crop enhancement for coming-generation husbandry. They punctuate how protein analysis helps understand factory growth, stress responses, and particularity development. The study emphasizes the eventuality of proteomics in enhancing crop adaptability, yield, and sustainability. Their findings contribute to developing innovative agrarian strategies for bettered productivity. Digra et al. (2024*)* bandy the integration of Green- IoT, and machine literacy in perfection husbandry, emphasizing their role in optimizing resource use, and automating husbandry processes. They punctuate advancements in smart husbandry technologies that enhance effectiveness, sustainability, and real-time decision tree. The study explores how IoT-driven data collection and AI-grounded analysis ameliorate crop monitoring and yield prediction.

Tran et al. (2024) explore cyclotides as coming-generation biopesticides for eco-friendly husbandry, emphasizing their eventuality as natural pest control agents. They punctuate cyclotides' stability, bioactivity, and environmental safety, making them sustainable volition to chemical fungicides. The study discusses their mode of action, effectiveness against pests, and implicit operations in integrated pest operation. Ray et al. (2024) examine the integration of Next Generation Science Standards (NGSS) into interdisciplinary husbandry, food, and natural coffers education. They stress the significance of combining scientific principles with agrarian education to foster invention and sustainability. The study emphasizes how NGSS can enhance scholars' critical thinking and problem- working.

Ding et al. (2024) provide a comprehensive check on coming-generation computer vision methods for factory complaint monitoring in perfection husbandry. They present a taxonomy of AI-driven styles, experimental perceptivity, and future exploration directions. The study highlights how deep literacy and image processing enhance early complaint discovery, and decision-making in husbandry. Bajpai et al. (2024) explore the development of coming-generation biostimulants for sustainable husbandry, emphasizing their role in enhancing factory growth, and stress adaptability. They punctuate advancements in bio-stimulant phrasings that ameliorate nutrient uptake, soil health, and crop productivity. The study discusses how biostimulants can reduce dependence on chemical diseases and fungicides. Their exploration underscores the significance of eco-friendly systems for long-term agrarian sustainability.

Arora and Mishra (2024) examine next-generation microbe-grounded bioinoculants, and their role in sustainable husbandry, and food security. They bandy how salutary microbes enhance soil fertility, factory health, and crop yield. The study highlights bio-inoculants' eventuality in replacing synthetic agrochemicals while promoting eco-friendly husbandry practices. Their findings emphasize the significance of microbial systems in addressing global agrarian challenges. Chahin et al. (2024) bandy weed discovery in smart husbandry using AI and 5G technology as part of the Smart Country Project. They punctuate how real-time data processing and machine literacy enhance perfection weed operation. The study emphasizes the role of high-speed connectivity in automating husbandry tasks. Their findings contribute to perfecting effectiveness and sustainability in ultramodern husbandry.

Dawar et al. (2024) explore the operation of resolvable AI (XAI) in coming-generation husbandry to ameliorate translucency in AI-driven husbandry opinions. They punctuate how XAI enhances trust, responsibility, and stoner relinquishment in smart husbandry systems. The study discusses current AI operations and future challenges in agrarian mechanization. Their exploration emphasizes the significance of interpretable AI for effective decision tree. Sahu et al. (2023) review environmental DNA (eDNA), an important instrument for biodiversity conservation, and species covering. They punctuate its capability to describe invasive species and assess ecosystem health through inheritable material analysis. The study emphasizes Edna's non-invasive approach in tracking biodiversity across different surroundings. Their findings contribute to advancing ecological exploration and justifiable conservancy master plans.

Verma et al. (2023) bandy the integration of AI with UAVs to enhance perfection husbandry by perfecting crop monitoring, and pest discovery. They punctuate UAV advancements in real-time data collection, and analysis for optimized husbandry. The study emphasizes the role of AI in resource operation and sustainability. Their findings contribute to developing intelligent, technology-driven agrarian systems. Bhatla et al. (2023) explore the part of unmixed biochemicals, and nanomaterials as coming-creation biostimulants for harvest rigidity under abiotic tension. They punctuate advancements in nanotechnology for perfecting nutrient uptake, and stress forbearance. The study discusses eco-friendly systems for enhancing factory growth and productivity. Their exploration contributes to sustainable husbandry through innovative bio-stimulant phrasings.

Joarder (2023) examines peasants' perspectives on involving the coming generation in husbandry in Bangladesh. The study highlights socio-profitable challenges, pastoral-to-civic migration, and policy gaps affecting youth participation. It discusses strategies for making tilling more seductive to youngish generations. Their findings emphasize the need for support systems to ensure agrarian sustainability. Tagkopoulos et al. (2022) punctuate the part of AI in advancing food products, force chains, and sustainability through the AI Institute for Next Generation Food networks. They emphasize interdisciplinary collaboration for addressing food security challenges. Their exploration contributes to AI integration in ultramodern husbandry.

Siponen et al. (2022) bandy a coming-generation task regulator using Optical Particle Counter (OPC) Unified Architecture to ameliorate robotization and effectiveness in smart husbandry. They punctuate their part in enhancing machine interoperability and perfection husbandry. Their study supports standardized communication for optimized ranch operation. Thomas et al. (2022) examine the use of Light Detection and Ranging (LiDAR) and radar technologies for perfecting dairy cattle feeding systems. They punctuate the benefits of real-time monitoring in optimizing feed distribution and reducing waste. Their exploration enhances mechanization in livestock nutrition and ranch effectiveness *a* real-time simulation and data-driven decision-making in husbandry. They punctuate its eventuality for perfecting effectiveness, reducing resource waste, and enhancing prophetic analytics. Their study contributes to coming-generation perfection husbandry technologies.

10.3 THE BUSINESS OF FRAGILE SENSING

10.3.1 Introduction to Remote Sensing Technologies

Remote sensing technologies, similar to satellite imagery and drone-grounded detectors, have revolutionized ultramodern husbandry by enabling growers to contain authentic data on crop wellness, soil stipulations, and environmental stipulants. Satellite imaging provides a broad, and detailed view of vast agrarian geographies, allowing growers to cover large-scale crop health, describe patterns in growth and identify anomalies that may indicate implicit issues similar to failure stress, pest infestations, or nutrient scarcities.

Drones, equipped with technical detectors, provide high-resolution imaging capabilities that offer more localized and precise data collection. Unlike satellite imagery, which may be affected by pall cover and lower resolution, drones can capture images with much lesser detail, and in near real-time. This allows for targeted analysis of specific sections of cropland, enabling early discovery of problems, and minimizing implicit damage.

The integration of remote-sensing technologies is essential for ultramodern perfection husbandry, as they allow growers to cover crop growth, and environmental changes more effectively. By exercising these advanced tools, growers can make data-driven opinions to optimize irrigation, fertilization, and pest regulator, eventually leading to better yields, reduced resource destruction and lesser sustainability in agrarian practices.

10.3.2 The Integration of AI in Agriculture

AI plays a transformative role in processing and assaying the vast quantities of data collected through remote sensing tools. Intelligent analytics apply machine learning techniques to detect patterns, forecast yields, and enhance farm management strategies, therefore enhancing the effectiveness of perfection husbandry.

For example, AI can reuse high-resolution drone imagery to describe early signs of pest infestations or nutrient scarcities, allowing growers to take timely corrective measures. This capability helps prevent minor issues from rising into major crop failures. AI-powered image analysis can also separate between different types of crop stress, distinguishing whether a factory is suffering from water failure, a fungal infection, or a soil nutrient imbalance, which is pivotal for applying precise interventions.

Likewise, AI integrates meteorological data to give prophetic recommendations on irrigation and fertilization schedules. By assaying rainfall vaccinations and soil humidity situations, AI systems can suggest optimal times for soddening crops, therefore reducing gratuitous water operation, and perfecting effectiveness. These intelligent systems contribute to reducing waste, adding productivity, and icing further sustainable agrarian practices, making AI a foundation of ultramodern perfection husbandry.

These technologies, as shown in Figure 10.1, leverage AI and remote sensing for tasks such as crop monitoring and yield prediction . Crop Monitoring and Health (30%) uses AI-powered drones and satellite imaging to describe factory stress, optimize irrigation, and ameliorate productivity. Pest and Disease Detection (25%) leverages AI-driven image recognition to identify infestations beforehand, reducing fungicide overuse, and crop damage. Soil & Irrigation Optimization (20%) relies on IoT detectors, and AI analytics to manage water, and nutrients efficiently, precluding waste. Yield Prediction and Decision Making (25%) utilizes machine literacy models to read crop affairs, helping growers plant crops, and reduce losses.

10.3.3 Operations of Remote Sensing and AI

Remote sensing and AI have different operations across colorful agrarian processes. These technologies permit real-time harvest monitoring, predictive analytics, and automation, steering to increased

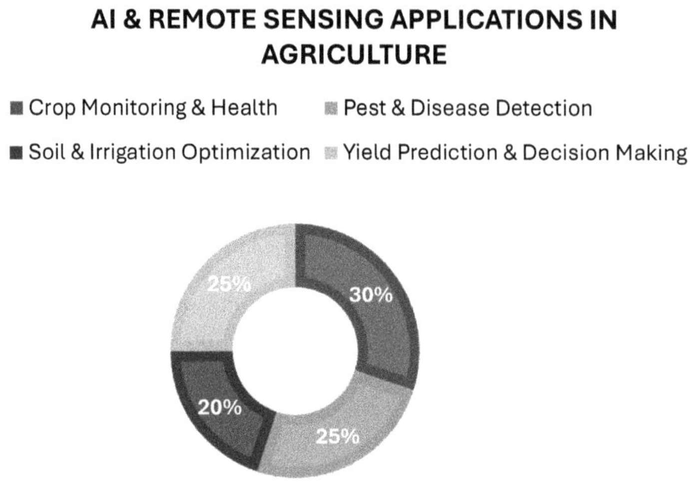

FIGURE 10.1 AI and remote sensing applications in agriculture, highlighting crop monitoring, pest detection, irrigation optimization, and yield prediction. (Created by the author, 2025.)

effectiveness, and downgraded costs in husbandry operations. One of the crucial operations is crop yield prediction, where AI algorithms dissect literal data, soil composition, and climate patterns to read crop labor. This allows growers to plan better for request demands and optimize their product cycles. Remote sensing is pivotal for detecting factory stress and conditions. By using multispectral and hyperspectral imaging, growers can identify stressed-out shops before symptoms become visible to the naked eye. AI-powered models also interpret these images to suggest corrective conduct, similar to targeted fungicide operation or soil emendations.

Another significant operation is irrigation optimization. Remote sensing data combined with AI analytics ensures that water coffers are used efficiently. By integrating satellite or drone images with on-ground soil detectors, AI systems determine the precise water conditions of different field sections, reducing water destruction and precluding over-irrigation. These technologies also aid in mapping soil fertility, helping growers apply diseases only where necessary, and reducing input costs, and environmental impact.

10.3.4 Benefits and Challenges of Precision Farming

Precision husbandry, enabled by AI and remote sensing technologies, offers multitudinous advantages, including better yields, reduced resource consumption, and enhanced sustainability. One of the primary benefits is increased agrarian productivity, as growers can optimize their resource application grounded on real-time data perceptivity. By applying diseases, fungicides, and irrigation only where demanded, perfect husbandry reduces input costs and minimizes environmental pollution.

Another crucial advantage is the early discovery of crop conditions and pest infestations. By exercising AI-driven analysis of remote-sensing images, growers can intermediate instantly, precluding wide crop damage and icing healthier crops. Perfection husbandry helps in soil conservation by minimizing the inordinate use of chemical inputs, thereby maintaining long-term soil fertility.

Still, despite its benefits, perfect husbandry faces several challenges. High original investment costs pose a significant hedge, especially for smallholder growers in developing regions. The demand for advanced structure, similar to high-speed internet connectivity, and pall-grounded computing, further limits availability. Data security enterprises arise with the adding reliance on AI-driven agrarian systems, challenging robust programs to cover growers' sensitive data. Addressing these challenges is essential for icing wide relinquishment and maximizing the eventuality of coming-generation agrarian technologies.

10.4 IOT APPLICATIONS AND SMART FARMING TECHNOLOGIES

10.4.1 The Role of the Internet in Agriculture

IoT refers to the connected network of smart biases that collect, exchange, and dissect data to automate and optimize colorful processes. In husbandry, IoT technologies play a pivotal part in enhancing productivity, resource operation, and effectiveness. IoT-enabled systems is similar to soil humidity detectors, temperature detectors, automated irrigation systems, and GPS-equipped ministry gives growers real-time data on crop fitness, downfall contingencies, and soil parcels. These perceptivity help optimize decision-making processes, allowing growers to respond instantly to changes in environmental factors. With IoT integration, growers can ever cover and control husbandry operations using mobile apps or pall-grounded platforms.

This digital metamorphosis of husbandry enables perfect husbandry, which minimizes resource destruction, improves yields, and ensures sustainable husbandry practices. For illustration, smart sensors track soil changes and automatically adjust irrigation schedules, ensuring efficient water use. Also, real-time monitoring of beast health through wearable IoT bias ensures timely medical interventions, reducing losses due to complaint outbreaks. Overall, IoT serves as the backbone of ultramodern smart husbandry, fostering data-driven decision-making that enhances overall ranch effectiveness, and sustainability.

10.4.2 Crucial IoT Operations in Farming

IoT operations in tilling span a wide range of functionalities, from simple data collection tools to complex independent systems that operate without mortal intervention. One of the most prominent operations is smart irrigation, where IoT-grounded humidity detectors cover soil conditions, and spark irrigation only when necessary. This minimizes water consumption while icing crops admit acceptable hydration. Moreover, automated rainfall stations give real-time meteorological data, enabling growers to prepare for adverse rainfall conditions, and acclimate their agrarian practices consequently.

Another crucial IoT operation is the use of smart ministry similar to independent tractors, robotic harvesters, and perfection agronomists. These machines are equipped with GPS, AI, and IoT connectivity to optimize sowing, fertilization, and harvesting. Drones fitted with high-resolution cameras, and multispectral detectors give upstanding imaging of granges, helping describe pest infestations, nutrient scarcities, and crop conditions. Likewise, IoT technology is employed in beast operation, where detectors track beast movement, body temperature, and feeding patterns, icing early discovery of health issues. IoT-powered force chain operation systems enhance translucency in food products by tracking yield from ranch to request.

Smart storehouse systems cover temperature, and moisture situations in storages, precluding post-harvest losses due to corruption. Through these advanced operations, IoT is revolutionizing husbandry by adding effectiveness, reducing labor reliance, and perfecting overall agrarian affairs.

10.4.3 Advantages of IoT in Smart Farming

The integration of IoT in smart husbandry offers multitudinous advantages, significantly perfecting effectiveness, sustainability, and profitability. One of the primary benefits is real-time monitoring of crops, soil qualifications, and environmental factors. Growers can search data through mobile operations, allowing them to frame good beliefs remarking irrigation, fertilization, and pest control. This visionary approach minimizes pitfalls associated with changeable rainfall and pest outbreaks, thereby reducing crop losses.

Another crucial advantage of IoT in husbandry is the optimization of resource applications. By precisely applying water, diseases, and fungicides grounded on real-time data, growers can reduce destruction and lower functional costs. Smart irrigation systems ensure that water is used efficiently, conserving this precious resource while maintaining optimal crop hydration. Targeted fungicide operation reduces chemical runoff, minimizing environmental pollution and perfecting soil health. IoT also enhances ranch outfit operation. Detectors embedded in machinery provide insights into machine performance, energy consumption, and conservation requirements. This prophetic conservation approach reduces unanticipated breakdowns, extends outfit lifetime, and minimizes time-out.

In beast husbandry, wearable IoT bias tracks the health, and movement of creatures, icing early complaint discovery and perfecting productivity. By integrating IoT technology, husbandry becomes further data-driven, reducing homemade labor conditions and adding productivity. As a result, growers witness advanced yields, reduced costs, and lesser sustainability, eventually contributing to global food security.

10.4.4 Expostulations and Future Lookouts

Despite the transformative eventuality of IoT in husbandry, several challenges hamper its wide relinquishment. One of the primary obstacles is the lack of dependable internet connectivity in pastoral and remote husbandry areas. Numerous IoT operations bear nonstop data transmission, and without high-speed internet, their effectiveness is compromised. Sweats to expand broadband access in agrarian regions are pivotal for maximizing the benefits of IoT technologies.

Another significant challenge is the high original investment needed for IoT relinquishment. The cost of copping, and installing IoT-enabled detectors, automated ministry, and pall-grounded data analytics platforms can be prohibitive for smallholder growers. Governments and agrarian associations must give fiscal support, subventions, and training programs to grease the relinquishment of smart husbandry technologies, icing indifferent access for all growers.

The adoption trends (Figure 10.2) illustrate how IoT and AI have been widely adopted due to their efficiency and decision-making capabilities. The graph shows that the relinquishment of agrarian technologies varies, with Precision Farming (80%) and IoT in Farming (75%) leading due to effectiveness and real-time monitoring. AI in Agriculture (65%) enhances decision-making with

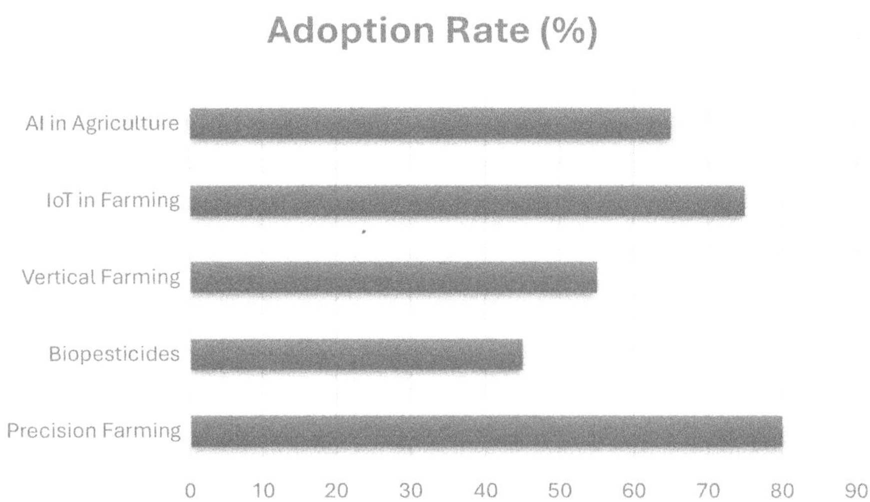

FIGURE 10.2 Adoption rates of agricultural technologies including precision farming, IoT, AI, and vertical farming. (Created by the author, 2025.)

prophetic analytics, while Vertical Farming (55%) is growing despite high costs. Bio-pesticides (45%) face lower relinquishment due to growers' preference for chemical druthers. Overall, high-tech systems are transubstantiating husbandry but cost-effective strategies are demanded for wider relinquishment.

Data security and sequestration enterprises also pose challenges to IoT relinquishment. Since IoT bias collects vast quantities of sensitive agrarian data, icing secure data storehouse, and protection against cyber pitfalls is essential. Growers need to be educated on cybersecurity practices to guard their data from unauthorized access.

Looking ahead, the future of IoT in husbandry appears promising. Developments in artificial intelligence, edge computing, and blockchain technology will refine IoT. The integration of these technologies enhances operations. As connectivity improves, and IoT bias becomes more affordable, indeed small-scale growers will be suitable to work smart husbandry systems. The nonstop elaboration of IoT-driven husbandry will contribute to sustainable food products, advanced effectiveness, and increased adaptability against climate change, icing a more robust agrarian sector for future generations.

10.5 VERTICAL FARMING, CONTROLLED-TERRAIN HUSBANDRY, AND URBAN FARMING IN THE NEXT GENERATION

10.5.1 Introduction to Next-Generation Vertical Farming

Vertical husbandry is set to play a more transformative role in the coming generation, using advancements in mechanization, AI, and biotechnology. The future of perpendicular husbandry will see completely automated systems where AI-powered robots would handle planting, covering, and harvesting crops with perfection. Smart sensors will continuously monitor plant health, adjusting nutrients, water, and lighting in sync with real-time needs to maximize yield and quality. Machine literacy algorithms will prognosticate and help conditions before they spread, icing harmonious and dependable product. Future perpendicular granges will decreasingly integrate renewable energy sources similar to solar panels, and wind turbines, reducing reliance on conventional power grids.

Inventions in LED lighting will ameliorate energy effectiveness, and pretend optimal sun conditions, further enhancing growth rates. Also, inheritable engineering and Clustered Regularly Interspaced Short Palindromic Repeats (CRISPR) technologies may play a role in developing crops specifically designed for perpendicular husbandry surroundings, optimizing them for rapid-fire growth, nutrient viscosity, and adaptability against pests. As the civic population rises, and pastoralist land becomes scarcer, coming-generation perpendicular husbandry will give a sustainable and largely effective result to meet food demands.

10.5.2 The Future of Controlled-Terrain Husbandry Controlled-Environment Agriculture (CEA)

The coming generation of CEA will see the full integration of smart technology, allowing for effective and climate-independent husbandry. AI-driven climate control systems will cover and regulate every aspect of the growing terrain, including temperature, moisture, CO_2 situations, and lighting. These automated systems will ensure optimal growing conditions all the time, allowing crops to thrive irrespective of external rainfall conditions. Future CEA will heavily use hydroponics, aeroponics, and aquaponics, drastically reducing water operation compared to traditional husbandry. Inventions in water recycling and desalination will make water use indeed more effective, ensuring that food products remain sustainable. Also, the perpetration of blockchain technology will give transparent shadowing of food from seed to trade, allowing consumers to corroborate the quality, and origin of their yield. As robotics and AI continue to evolve, CEA will become decreasingly independent, minimizing the need for human intervention.

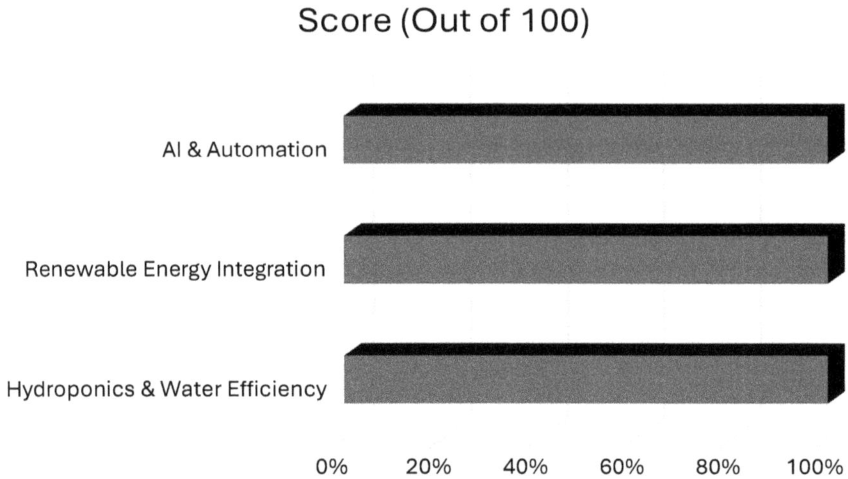

FIGURE 10.3 Key technological advancements in next-generation agriculture, such as AI, hydroponics, and blockchain. (Created by the author, 2025.)

Figure 10.3 highlights crucial advancements in coming-generation husbandry. AI and automation (90/100) play a pivotal role in effectiveness, while renewable energy (80/100) reduces reliance on traditional power sources. Hydroponics and water effectiveness (85/100) ensure sustainable water use. The illustration highlights key advancements in next-generation agriculture. AI and automation (90/100) play a pivotal role in effectiveness, while renewable energy (80/100) reduces reliance on traditional power sources. Hydroponics and water efficiency (85/100) ensure sustainable water use, as shown in Figure 10.3 visualizes the synergy between technologies like AI, hydroponics, and blockchain in modern agriculture (Author, 2025). Overall, these technologies drive sustainability and effectiveness in ultramodern husbandry.

Drones and robotic arms will manage planting, nutrient distribution, and harvesting, making tilling more effective and cost-effective. Also, future CEA systems will integrate with renewable energy sources and civic structure, turning underutilized spaces similar to abandoned storages or indeed underground cellarages into largely productive granges.

10.5.3 Next-Generation Urban and Its Expanding Part

Civic husbandry in the coming generation will be a crucial element of flexible and tone-sufficient metropolises. As civic populations continue to grow, metropolises will integrate husbandry directly into domestic and marketable structures. Future civic granges will be bedded into towers, rooftops, and community capitals, reducing the need for expansive transportation, and icing that fresh, locally-grown food is available all the time.

Advancements in smart civic husbandry technologies will allow for flawless integration with megacity structures. Automated rooftop glasshouses will acclimate their climate grounded on AI-anatomized rainfall prognostications, optimizing conditions for crop growth. Vertical auditoriums and hydroponic walls will become common in high-rise structures, transforming cityscapes into lush, food-producing ecosystems. Also, civic granges will integrate with IoT networks, allowing residers to cover and, indeed, share in food products through mobile operations.

Beyond food products, coming-generation civic granges will contribute to better air quality, increased biodiversity, and the reduction of civic heat islets. By combining civic husbandry with renewable energy sources, rainwater harvesting, and waste recycling, metropolises will become more sustainable and less dependent on external food force chains, creating a decentralized and flexible food system.

10.5.4 Terrace Farming

Sundeck husbandry in the coming generation will suffer a significant metamorphosis, integrating colorful practices to enhance productivity and sustainability. Hydroponic and aeroponic systems will minimize water operation, while smart rainwater harvesting and bioengineered soil stabilizers will help corrosion, and ameliorate soil fertility. To maximize space, perpendicular husbandry methods similar to piled hydroponic setups, modular aeroponic halls, and LED-optimized multi-layered glasshouses will be incorporated, allowing for increased crop product in limited areas. Sustainability will be further enhanced by integrating renewable energy sources similar to solar panels, wind turbines, and micro-hydroelectric creators, reducing reliance on external power grids. Climate-smart crops developed through CRISPR gene-editing will be optimized for steep terrains, icing adaptability against extreme rainfall and perfecting nutrient viscosity.

AI-powered independent husbandry outfits, including robotic arms, tone-driving micro-tractors, and automated harvesting drones, will replace traditional labor-intensive processes, perfecting effectiveness and reducing costs. Blockchain-grounded digital husbandry sidekicks will help growers cover crop growth, prognosticate yields, and manage force chains with lesser translucency. As urbanization increases, sundeck husbandry will expand beyond pastoral areas into smart rooftop granges, and hydroponic perpendicular walls in high-rise structures, integrating AI-driven climate control systems for decentralized civic food products.

Despite challenges similar to high original investment costs, and structure conditions, advancements in cost-effective agri-tech systems, government impulses, and private sector investments will drive large-scale relinquishment. Eventually, coming-generation sundeck husbandry will combine slice-edge technology, sustainable resource operation, and mechanization to produce a largely effective, climate-flexible, and productive agrarian system, icing food security for future generations.

10.5.5 Challenges and Sustainability Considerations for the Future

Despite promising advancements in perpendicular husbandry, controlled-terrain husbandry, and civic husbandry, significant challenges remain for the coming generation. High energy consumption remains a concern, particularly for perpendicular granges that are calculated on artificial lighting. Still, improvements in solar energy, battery storehouses, and energy-efficient LEDs will help alleviate this issue. Governments and private sector hookups will play a pivotal role in backing and incentivizing green energy integration within agrarian systems. Another challenge is scalability. While perpendicular and civic granges are effective in high-viscosity areas, they may struggle to match the large-scale product of traditional husbandry. Future exploration and investment will concentrate on developing large-scale CEA capitals capable of mass-producing chief crops similar to wheat, and rice in addition to lush flora, and vegetables.

Data security and sequestration enterprises will arise as AI-driven husbandry collects massive quantities of data on food products. Transparent data programs and blockchain-grounded shadowing systems will be necessary to ensure the ethical operation of agrarian data. Overall, the coming generation of husbandry technologies will drive sustainability, effectiveness, and adaptability in food products. With continued advancements in AI, robotics, and green energy, perpendicular husbandry, CEA, and civic husbandry will revise how metropolises and communities grow and consume food, creating a more sustainable and food-secure future.

Climate adaptability will become a critical focus, as changing rainfall patterns and environmental insecurity place lesser pressure on food systems worldwide. Advanced covering systems, prophetic analytics, and climate-adaptive crop kinds will be essential to ensure stable yields under changeable conditions. Water operation will also pose a patient challenge, especially in water failure regions. The integration of perfect irrigation technologies, water recycling systems, and failure-resistant crops will help optimize water operation and minimize waste. Fostering public trust and consumer acceptance of technologically advanced husbandry practices will bear clear communication, translucency, and education enterprise. Bridging the gap between traditional agrarian values and rising inventions will be crucial to icing broad societal support for these transformative systems.

Transnational collaboration and knowledge sharing will further accelerate progress, as countries adopt sustainable practices, technology-driven systems, and policy frameworks to treat overall food protection demurs inclusively. Eventually, the confluence of sustainable technologies, cooperative programs, and responsible data stewardship will shape the future of husbandry, ensuring that both current and future generations profit from flexible, effective, and environmentally conscious food systems. To further enhance the inclusivity of future husbandry systems, targeted programs will be necessary to ensure that smallholder growers and pastoral communities have access to coming generation technologies, fiscal support, and capacity-structure programs. Without indifferent access, there's a threat that advanced agrarian systems could consolidate the digital peak, concentrating benefits in technologically advanced regions while marginalizing lower, resource-constrained directors.

The long-term sustainability of rising agrarian systems will also depend on conserving soil health and biodiversity within controlled and civic husbandry surroundings. Integrating soil microbiome exploration, crop diversification strategies, and bioinoculants into perpendicular husbandry, and CEA systems will help maintain ecological balance, and adaptability indeed in largely controlled settings. In addition to technological advancements, holistic policy fabrics will be needed to harmonize agrarian invention with broader sustainability pretensions, similar to climate mitigation, pastoral development, and food justice. Coordinated sweats across sectors including energy, water operation, civic planning, and education will be essential to creating agrarian ecosystems that support both environmental health and human well-being.

Ethical considerations will also play a decreasingly important role, particularly regarding inheritable revision, data sequestration, and commercial control over critical agrarian technologies. Icing that ethical guidelines and social impact assessments are integrated into technological development processes will help balance invention with the public interest. Eventually, flexible food systems will need to incorporate community-position participation, ensuring that technological advancements align with original societies, knowledge systems, and salutary preferences.

10.6 SUSTAINABLE PEST MANAGEMENT THROUGH BIOPESTICIDES AND BIOINOCULANTS IN THE NEXT GENERATION

10.6.1 Cattle Rearing

The elaboration of cattle rearing in coming-generation husbandry will be driven by AI-powered monitoring, perfect nutrition, automated complaint prevention, sustainable feed druthers, and renewable energy integration. The integration of IoT-enabled smart collars, AI-powered drones, and real-time biometric sensors will enable continuous monitoring of cattle health, movement, and feeding behavior, icing early complaint discovery and optimizing herd operation. Automated feeding stations will use AI-driven analytics to apportion precise nutrient phrasings predicated on a beast's age, weight, and product cycle, enhancing growth rates while reducing feed waste. To minimize the environmental footprint of beast husbandry, lab-grown feed, seaweed- predicated methane-reducing supplements, and hydroponic fodder systems will replace conventional grain-fed diets, significantly reducing land declination, and hothouse gas emigrations. CRISPR-predicated heritable advancements will enhance complaint resistance, ameliorate reduplication rates, and ensure advanced-quality meat and dairy products.

Inventions in automated abuse systems, AI-powered meat processing, and blockchain-integrated force chains will enhance effectiveness, hygiene, and traceability, ensuring that dairy and meat products meet strict safety and sustainability guidelines. AI-driven complaint prophecy models and robotic vaccine administration systems will enable visionary health operations, reducing the spread of infections while lowering the need for antibiotics. Sustainability will be further corroborated by solar-powered barns, biogas digesters for excreta to energy conversion, and AI-controlled climate regulation systems, which will ensure optimal environmental conditions while reducing functional costs.

Despite the promising advancements, challenges analogous to high original investment costs, nonsupervisory complications, and farmer adaptation to AI-driven systems will need to be addressed through government impulses, private-sector backing, and digital training programs. Still, as global demand for sustainably produced meat and dairy continues to rise, next-generation cattle husbandry will crop as a largely effective, climate-flexible, and technologically advanced sedulity, balancing productivity with environmental responsibility.

Next-generation cattle rearing will incorporate indirect waste operation systems where advanced anaerobic digesters convert ordure into biogas to power ranch operations, with nutrient-rich digestate repurposed as an organic toxin. Composting systems will be further optimized to enhance nutrient recovery and reduce emigration. Digital binary technology will enable virtual modeling of individual creatures, and herds, allowing prophetic simulations for health interventions, feeding strategies, and environmental changes to optimize productivity and resource effectiveness. Low-emigration ordure operation, combined with carbon prisoner and soil enrichment technologies, will support carbon-neutral or carbon-negative cattle tilling.

Regenerative grazing systems will enhance soil carbon insulation, biodiversity, and pasturage health, while climate-adaptive sanctum designs and feed phrasings will ameliorate adaptability in the face of extreme rainfall events. Co-operative invention ecosystems involving public–private hookups, planter cooperatives, and digital training programs will be critical in driving technological relinquishment and fostering nonstop enhancement. Though original investment costs, nonsupervisory challenges, and the need for planter adaption may decelerate relinquishment, these inventions will inclusively transfigure cattle tilling into a largely effective, climate-flexible, and environmentally responsible sector able to meet rising global demand for sustainable meat and dairy.

10.6.2 The Problem with Traditional Fungicides

In the coming generation, the challenges associated with traditional fungicides will become indeed more pronounced. Due to adding pest resistance, the effectiveness of chemical fungicides will decline, causing growers to use advanced boluses, which will further complicate environmental damage. Regulatory agencies worldwide will probably put stricter restrictions on chemical fungicide operations due to enterprises over soil declination, water impurity, and biodiversity loss. Climate change will also consolidate the need for indispensable pest operation systems, as rising temperatures and changeable rainfall patterns will expand the range of pests, and conditions affecting crops.

Future agrarian systems will concentrate on integrated pest operation Integrated Pest Management (IPM) strategies that combine natural control styles, perfect operation technologies, and inheritable advancements. Digital platforms and AI-driven systems will cover pest outbreaks in real time, allowing for further targeted and sustainable interventions. The coming generation of growers will calculate more on data analytics, drone surveillance, and detector-grounded monitoring to reduce reliance on dangerous chemical fungicides.

10.6.3 Biopesticides – A Sustainable and Indispensable Solution

The coming generation of biopesticides will be more advanced, with better efficacy, longer shelf life, and enhanced targeting mechanisms. Future innovations will focus on tailored microbial biopesticides, engineered to target specific pests without harming beneficial organisms. With the advancement of nanotechnology, biopesticide phrasings will become more stable and effective, allowing for precise delivery to affected areas, reducing waste, and maximizing effectiveness.

AI-powered analytics and smart husbandry techniques will help determine the most effective biopesticide operations grounded on environmental conditions, soil health, and pest geste. Biopesticides will be integrated with smart delivery systems similar to automated sprayers, and upstanding drones, icing that crops confess the precise quantum of guard at the accurate occasion. Governments and private enterprises will probably invest more in exploration, and products to make biopesticides more affordable, and accessible to growers worldwide.

10.6.4 Bioinoculants and Their Part in Soil Health

In the coming generation, bioinoculants will play a pivotal part in restoring and maintaining soil health. As agrarian lands continue to suffer from chemical overuse, there will be a growing shift toward regenerative husbandry practices that incorporate bioinoculants to ameliorate soil microbiomes. Advanced formulations will be designed to work in different soil types, and climatic conditions, enhancing crop adaptability against environmental stresses similar to failure, extreme temperatures, and poor soil fertility. Future bioinoculants will probably be combined with AI-driven soil analysis tools to give real-time recommendations for their operation. Smart husbandry technologies, including Internet of Things (IoT) enabled soil detectors, will describe nutrient scarcities, and automatically administer the applicable bioinoculants to enhance soil vitality. Also, coming-generation bioinoculants will be integrated with synthetic biology inventions, allowing scientists to produce custom-designed microbial strains that enhance factory impunity, and increase nutrient immersion effectiveness.

As depicted in Figure 10.4, biopesticides and bioinoculants are emerging as leading sustainable pest control solutions (Author, 2025). The pie map illustrates the distribution of sustainable pest operation methods, pressing the shift toward eco-friendly systems. Biopesticides account for the

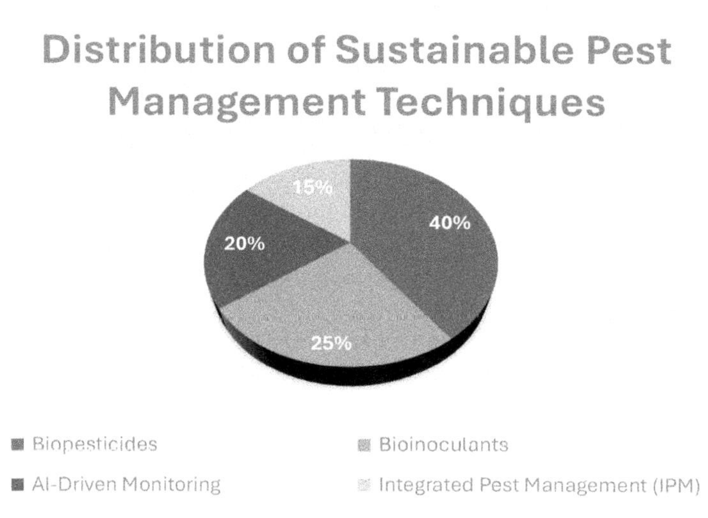

FIGURE 10.4 Distribution of pest management techniques. (Created by the author, 2025.)

largest share at 40%, offering an effective volition to chemical fungicides while minimizing environmental detriment. Bioinoculants make up 25%, playing a pivotal part in enhancing soil health and boosting factory impunity. AI-driven monitoring contributes 20%, exercising AI and IoT-grounded systems to track pest populations and optimize fungicide operations. Incipiently, Integrated Pest Management (IPM) holds 15%, combining colorful sustainable styles to ensure long-term and effective pest control. These techniques inclusively represent the future of environmentally responsible husbandry.

10.6.5 Challenges in the Relinquishment of Biopesticides and Bioinoculants

While the benefits of biopesticides, and bioinoculants are clear, challenges in their wide relinquishment will persist in the coming generation. The high cost of product and limited vacuity will remain walls, particularly for smallholder growers in developing regions. Still, as biotechnology advances and products scale up, prices are anticipated to decline, making these sustainable systems more accessible. Regulatory fabrics will also evolve to accommodate the rapid-fire development of biopesticides, and bioinoculants. Governments may introduce presto-track blessing processes, and incentivize sustainable husbandry practices through subventions, and duty benefits. Also, consumer demand for organically grown food will continue to rise, encouraging agribusinesses to invest in eco-friendly pest operation systems.

Educational programs and digital platforms will be necessary to bridge the knowledge gap. AI-driven agrarian premonitory systems will give growers real-time perceptivity on how to effectively apply biopesticides and bioinoculants. Public–private hookups will also play a crucial role in promoting exploration, mindfulness, and availability of sustainable pest operation methods.

10.7 CHALLENGES IN NEXT-GENERATION AGRICULTURE DIGITAL INEQUALITY, POLICY FRAMEWORKS, AND FUTURE PROSPECTS

10.7.1 Digital Inequality in Agriculture

As husbandry becomes further technology-driven, digital inequality remains one of the most burning challenges in coming-generation husbandry. Numerous growers, particularly in developing nations, and pastoral areas, warrant access to high-speed internet, smart bias, and essential digital structure. Without these coffers, smallholder growers struggle to borrow perfection husbandry methods, AI-powered analytics, or IoT-grounded monitoring systems, limiting their capability to ameliorate productivity and sustainability.

The digital peak is not only about internet access but also extends to the affordability and vacuity of digital tools. Advanced agrarian technologies, similar to satellite imagery, remote sensing bias, and AI-powered ranch operation software, bear significant investment, making them out of reach for small-scale growers. Also, a lack of digital knowledge further widens this gap, as numerous growers are unfamiliar with how to use these new tools effectively.

Bridging digital inequality in husbandry will bear coordinated sweat from governments, private companies, and governmental associations. Expanding rural broadband, subsidizing digital farming tools, and providing farmer-focused digital literacy programs will be essential. Mobile-grounded systems and AI-driven premonitory platforms acclimatized for low-resource settings could help homogenize access to coming-generation agrarian inventions. Ensuring that smallholder farmers can benefit from technological advancements is essential as next-generation agriculture integrates advanced technologies.

Table 10.1 compares the digital infrastructure between developing and developed regions, showing a significant gap in access and usage (Author, 2025). The table highlights the digital peak in husbandry, showing that developed countries (85) have advanced internet access, digital knowledge (80), and smart husbandry relinquishment (75) compared to developing nations (40, 35, and

TABLE 10.1
Comparison of Digital Infrastructure and Technology Adoption between Developed and Developing Agricultural Regions

Factor	Developed Countries (%)	Developing Countries (%)
Internet access	85	40
Digital literacy	80	35
SMART Farming tech adoption	75	25

Source: Created by the author (2025).

25 independently). Limited connectivity, education, and affordability hamper growers in developing regions from using AI-driven analytics, IoT-grounded irrigation, and automated tools. Bridging this gap requires investments in pastoral broadband, digital training, and cost-effective smart husbandry systems to ensure inclusive agrarian progress.

10.7.2 Policy Fabrics for Agricultural Innovation

The successful relinquishment of coming-generation agrarian technologies depends on robust policy fabrics that regulate and promote invention while icing sustainability, safety, and indifferent access. Governments play a critical role in shaping agrarian policy by setting guidelines for the use of genetically modified organisms (GMOs), biotechnology, perfection husbandry, and synthetic biology in food products. Programs must balance invention with ethical, and environmental considerations, ensuring that new technologies don't pose pitfalls to mortal health or biodiversity.

Also, policy fabrics should laboriously support the development of digital structure. This includes backing for exploration in agritech, duty impulses for growers espousing smart technologies, and regulations that encourage sustainable husbandry practices. Programs promoting open data access in husbandry where growers can partake and find real-time data on rainfall, soil health, and pest control can further accelerate invention and ameliorate decisions tree.

Moreover, transnational cooperation will be necessary to produce standardized regulations, and stylish practices for rising technologies. Issues similar to across-border data sharing, ethical AI operation in husbandry, and the regulation of lab-grown food will bear global agreement. Governments must also ensure that smallholder growers and marginalized communities aren't left before in the shift toward digital and biotech-driven husbandry by offering targeted fiscal and educational support.

10.7.3 Challenges in Technology Adoption

While coming-generation agrarian technologies offer immense eventuality, their relinquishment faces several obstacles. The high original cost of acquiring smart husbandry outfits, similar to AI-powered drones, IoT detectors, and automated irrigation systems, makes them inapproachable to numerous growers, particularly in developing nations. The lack of fiscal coffers, coupled with limited access to agrarian credit or government subventions, further complicates the relinquishment process.

Another major challenge is the lack of acceptable training, and specialized knowledge needed to operate, and maintain these advanced systems. Numerous growers are reluctant to borrow new technologies due to ignorance of digital tools, and enterprises' over trustability. Without proper training programs, and ongoing specialized support, indeed those who invest in ultramodern husbandry systems may struggle to use them effectively.

Likewise, nonsupervisory walls can decelerate the relinquishment of agrarian inventions. In some regions, strict blessing processes for biopesticides, Genetically Modified GM crops, or perfection-husbandry technologies produce detainments in request entry. Misinformation and dubitation about arising technologies, similar to AI-grounded ranch operation systems or gene-edited crops, also contribute to slow relinquishment rates. Prostrating these challenges will bear collaboration between technology providers, policymakers, and agrarian institutions to produce affordable, stoner-friendly, and well-supported systems that feed growers of all scales.

10.7.4 Future Prospects for Agricultural Technologies

The future of husbandry will be shaped by the uninterrupted integration of advanced technologies aimed at enhancing productivity, sustainability, and climate adaptability. AI-driven analytics, blockchain for food traceability, independent husbandry outfits, and gene-editing tools like CRISPR will revise food products. Precision husbandry will become the norm, with growers using AI-powered drones, robotic harvesters, and IoT-enabled soil monitoring to maximize effectiveness and minimize waste.

One of the most promising developments will be the wide relinquishment of regenerative husbandry, where technology is used to enhance soil health, biodiversity, and carbon insulation. Smart detectors and AI models will help growers transition to further sustainable husbandry styles by assaying soil microbiomes, optimizing crop gyration strategies, and reducing reliance on chemical inputs. Vertical husbandry and CEA will also see significant growth, particularly in civic areas, reducing reliance on traditional land-intensive agriculture.

Still, for these technologies to be truly transformative, they must be made popular and accessible to all farmers, irrespective of their profitable status or geographical position. Investment in digital structure, open-source agrarian data platforms, and education will be essential to ensuring that smallholder growers can share in and benefit from the coming agrarian revolution.

Transnational collaboration both in exploration and policy-timber will play a pivotal role in prostrating being walls. Cross-border enterprise for knowledge-sharing, technology transfer, and sustainability-concentrated investments will help produce a further inclusive and flexible global food system. In the coming decades, the integration of biotechnology, AI, and perfection husbandry will not only enhance food security but also contribute to reducing husbandry's environmental footmark, paving the arterial for a spruce, luxuriant, and more indifferent agricultural future.

10.8 NEED FOR NEXT-GENERATION AGRICULTURE

10.8.1 Integrating Technology

Agriculture is the backbone of human civilization, but with rapid-fire global changes, traditional husbandry styles are no longer sufficient to meet rising demands. The adding population, climate change, shrinking natural coffers, environmental enterprises, and food security issues bear a transition to coming-generation husbandry – a system that integrates technology, sustainability, and effectiveness to produce a flexible and productive agrarian sector. Next-generation husbandry will integrate indirect frugality principles, converting waste into biofertilizers, and bioenergy to close resource circles. Hyperactive original rainfall stations and prophetic climate analytics will enable real-time adaptations to planting, irrigation, and pest control, enhancing climate adaptability.

Mobile premonitory platforms and affordable IoT systems will ameliorate digital inclusivity for smallholders, icing indifferent access to advanced tools. AI-driven regenerative practices, perfect rotational grazing, and soil microbiome improvement will restore soil health, and boost biodiversity. Advanced breeding using gene editing and phenotyping will accelerate climate-flexible crop development. Carbon and biodiversity monitoring through remote sensing and blockchain will track environmental performance and unlock green backing. AI-powered force chains will reduce post-harvest losses by aligning the product with real-time demand, supported by adaptive logistics. Clear policy fabrics and targeted impulses will drive relinquishment and invention, and icing technology serves both productivity and sustainability goals.

The pictorial representation in Figure 10.5 for pivotal agricultural performance criteria for current husbandry practices scored out of 100. Yield effectiveness is moderate at 60, indicating decent productivity but room for improvement. Water operation effectiveness scores 50, suggesting a need for better irrigation and conservation styles. Chemical operation is high at 80, which is undesirable as devilish reliance on chemicals can harm the terrain and soil health. The automation position is low at 30, pressing minimal use of smart technologies and automation. Initially, the sustainability score stands at 40, reflecting challenges in eco-friendly husbandry. These perceptions suggest that espousing perfection husbandry, IoT, and AI could ameliorate effectiveness, reduce chemical reliance, and enhance sustainability. Performance scores in Figure 10.5 reflect how automation and sustainability remain key challenges in present-day agriculture (Author, 2025).

10.8.2 Growing Population Demands

The world population is anticipated to reach 10 billion by 2050, significantly adding to the requisition for food. Current husbandry techniques struggle to meet this demand, leading to enterprises about food deaths, and price affectation. Conventional styles calculate heavily on land expansion, but pastoralist land is limited, challenging a productivity-driven approach rather than an expansion-driven one.

Coming-generation husbandry focuses on yield optimization through technology. Precision husbandry, perpendicular husbandry, hydroponics, and geoponics offer results by adding productivity without taking fresh land. Inheritable revision (GMOs) and CRISPR gene-editing techniques can produce high-yield, pest-resistant crops that grow briskly, and more efficiently. Likewise, AI-driven agrarian planning helps growers dissect soil conditions, rainfall patterns, and crop health to make data-driven opinions that maximize yield while minimizing waste.

With population growth, urbanization also increases, limiting cropland vacuity. Civic husbandry, rooftop auditoriums, and CEA offer sustainable systems by integrating food products within metropolises, reducing transportation costs, and icing fresh yield vacuity for civic populations. Addressing future food demands will also bear diversification of food sources beyond traditional crops and

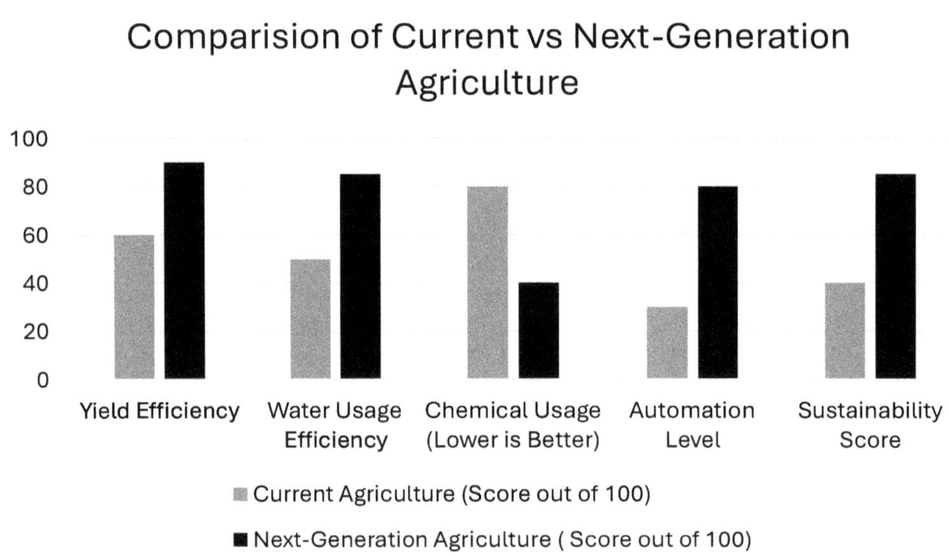

FIGURE 10.5 Comparison of current vs. next-generation agriculture. (Created by the author, 2025.)

beasts. Indispensable proteins are similar to lab-grown proteins in their composition and functionality , nonentity husbandry, and algae-grounded foods can round conventional food systems, reducing pressure on land, and water coffers while offering nutritional druthers. Advanced post-harvest operation systems, powered by real-time logistics tracking and cold chain optimization, will help minimize food loss across force chains, icing that increased product translates directly into better food vacuity. Decentralized food systems, supported by the community-driven food capitals and civic-pastoral hookups, will enhance Indigenous food security and adaptability to supply chain dislocations.

The projected surge in food demand shown in Figure 10.6 emphasizes the urgency of technology adoption (Author, 2025). The graph represents projected global food demand about population growth from 2030 to 2070. As the population increases, the demand for food rises proportionally. The trend highlights the challenge of sustaining food products to meet future requirements. Inventions in husbandry, similar to perfection husbandry, perpendicular husbandry, and biotechnology, will be pivotal in addressing this growing demand while optimizing coffers and minimizing environmental impact.

10.8.3 Climate Change Challenges

Husbandry is largely sensitive to climate change. Rising temperatures, irregular downfall patterns, and extreme rainfall events like famines, cataracts, and hurricanes disrupt traditional husbandry cycles, making crop failures more frequent. Traditional rain-fed husbandry styles are getting unreliable, and the unpredictability of climate patterns poses a severe threat to global food products. Coming-generation husbandry addresses climate risks through smart rainfall soothsaying, AI-grounded climate adaption strategies, and failure-resistant crops. Smart irrigation systems ensure optimal water operation, while detector-grounded soil monitoring ensures that crops admit precise nutrients, reducing water destruction, and nutrient runoff.

Likewise, climate-flexible crops are being developed using biotechnology to repel harsh environmental conditions. For illustration, scientists are negotiating heat-tolerant wheat and failure-resistant sludge to ensure stable yields under extreme temperatures. AI-powered husbandry models prognosticate forthcoming climate variations, helping growers acclimatize their planting, and harvesting schedules consequently. In addition to crop adaption, beast systems are

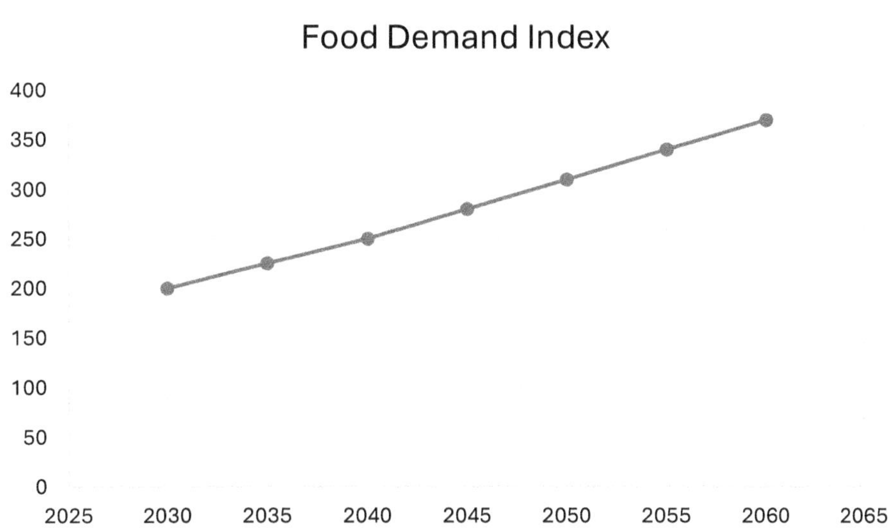

FIGURE 10.6 Food demand in upcoming generations. (Created by the author, 2025.)

also being reimagined to repel climate stress. Climate-flexible beast types, able to thrive in hotter, drier surroundings, are being developed through picky parentage and genomic selection methods. Carbon-smart husbandry practices, similar to integrating agroforestry, cover cropping and conservation tillage, enhance soil carbon insulation, perfect soil health, and boost its adaptability to extreme rainfall.

Microclimate operation systems, including shade-net husbandry, floating granges, and climate-controlled glasshouses, are arising to guard crops from extreme heat, storms, and irregular downfall. Also, parametric climate insurance programs are being introduced to help growers recover financially from climate-convince losses, offering payouts touched off by predefined rainfall events. Indirect resource use, where ranch waste is repurposed into bioenergy, compost, or beast feed, further enhances adaptability by reducing input reliance and structuring localized, tone-sufficient ranch ecosystems. By combining adaptive technologies, threat mitigation strategies, and flexible crop-beast systems, coming-generation husbandry aims to not only survive but also thrive amid adding climate query.

10.8.4 Resource Failure

Water and rich soil are finite coffers, yet traditional husbandry practices are largely resource-intensive. Conventional irrigation practices waste billions of liters of water annually, while overuse of plant diseases depletes soil fertility, leading to desertification, and reduced productivity over time. Coming-generation husbandry adopts perfect resource operation using IoT-enabled irrigation systems, regenerative husbandry, and indispensable growing methods. Drip irrigation ensures that every drop of water is used efficiently, significantly reducing waste.

Hydroponics and aeroponics allow crops to grow with 90 lower water than soil-grounded styles. Regenerative husbandry, which includes crop gyration, organic diseases, and reduced tillage, restores soil fertility and enhances biodiversity. AI-grounded ranch operation systems help track water operation, optimize planting patterns, and regulate soil health, making tilling further sustainable and lower resource-dependent.

10.8.5 Environmental Impact

Traditional husbandry is one of the largest contributors to deforestation, carbon emigration, and ecosystem destruction. Clearing timbers for husbandry leads to niche loss and increased hothouse gas emigrations. The inordinate use of chemical diseases and fungicides pollutes soil, and water bodies, harming both human health and biodiversity. Coming-generation husbandry promotes sustainable and eco-friendly approaches through organic husbandry, agroforestry, carbon insulation, and biopesticides.

By replacing synthetic chemicals with -based druthers, growers can reduce pollution while maintaining soil health. Agroforestry where trees and crops are grown together helps absorb carbon dioxide, restore ecosystems, and ameliorate ranch adaptability. Likewise, AI-driven sustainability models dissect carbon vestiges and give recommendations for low-emigration husbandry. Smart energy systems like solar-powered irrigation systems and electric husbandry outfits reduce dependence on fossil energies, making husbandry greener and more energy-effective.

10.8.6 Food Security and Nutrition

Food security is more than just adding products; it also means furnishing nutritional and accessible food for all. Malnutrition remains a major issue in numerous corridors of the world, with nutrient-deficient diets contributing to health problems similar to suppressed growth, weakened vulnerable systems, and habitual conditions. Coming-generation husbandry tackles this through biofortification, AI-driven food distribution, and advanced food storehouse results. Biofortified crops, similar to iron-rich sap, and vitamin A-enriched rice, give essential nutrients to communities that rely on staple foods.

AI-driven logistics help reduce food waste by optimizing force chains, ensuring food reaches consumers before corruption. Smart storehouse systems use IoT detectors to cover moisture, temperature, and corruption pitfalls, precluding post-harvest losses. Cold chain logistics and blockchain-grounded food tracking further ameliorate the effectiveness and translucency of food distribution, icing that nutritional food is available where it is demanded the most.

10.8.7 Relinquishment of Robotization and Robotics

Labor deaths are becoming a significant challenge in husbandry, especially as young generations shift down from tilling careers. Next-generation husbandry integrates mechanization robotics and AI to reduce dependence on homemade labor. Autonomous tractors, robotic harvesters, drone-grounded crop monitoring, and AI-powered weed discovery ameliorate effectiveness while reducing labor costs. Drones equipped with multispectral cameras overlook vast fields for early signs of pest infestations, soil scarcities, or crop conditions, allowing for targeted interventions rather than wide fungicide use.

Robotic arms can gather delicate fruits and vegetables with perfection, reducing damage and adding effectiveness. These inventions not only increase productivity but also make tilling a more seductive career option by incorporating technology-driven systems. Multi-crop adaptable robotic systems are arising, able to conform their tools, detectors, and algorithms to work across different crop types, planting patterns, and seasonal variations. These adaptable robots reduce the need for technical outfits, lowering capital costs for diversified granges. Swarm robotics, where lines of lower, cooperative robots perform coordinated tasks similar to sowing, weeding, and monitoring, offers a scalable volition for granges of varying sizes. By distributing tasks across multiple lower units, mass systems ameliorate redundancy, rigidity, and energy effectiveness.

Autonomous microclimate operation systems are being paired with robotics to stoutly acclimate environmental parameters within glasshouses and controlled surroundings. These systems, incorporating perfect irrigation, ventilation, and shading, work in musicale with robotic crop care to optimize growth conditions with minimum mortal intervention. The integration of real-time soil health monitoring with independent ministry allows robots to stoutly acclimate planting depth, fertilization rates, and irrigation volumes grounded on hyperactive-original soil conditions, maximizing resource use effectiveness.

Mobile robotic nurseries were able to nurture seedlings in optimized conditions before broadcasting them into fields to further enhance survival rates and reduce early-stage crop losses, particularly in grueling climates. The socio-economic impact of wide automation also presents both opportunities and challenges. While reducing labor deaths, these systems bear up-skilling pastoral workforces to operate, maintain, and repair high-tech outfits. Government programs, cooperatives, and agritech training centers will play a vital role in icing indifferent access to training and technology, particularly for smallholder growers. Ethical and nonsupervisory fabrics will need to evolve to address data power, liability for machine crimes, and the broader impacts of mechanization on pastoral employment and profitable stability. Developing programs that balance technological invention with social equity will be critical to icing inclusive agrarian metamorphosis.

Eventually, as climate change increases variability in growing conditions, climate-flexible independent systems able to conforming operations grounded on real-time rainfall vaticinators, soil humidity data, and pest pressure models will become necessary for maintaining stable yields, and ranch profitability in the face of the environmental query.

Figure 10.7 highlights the upward trend in AI and remote sensing technologies over eight decades (Author, 2025). This graph shows the relinquishment trends of remote sensing and AI in

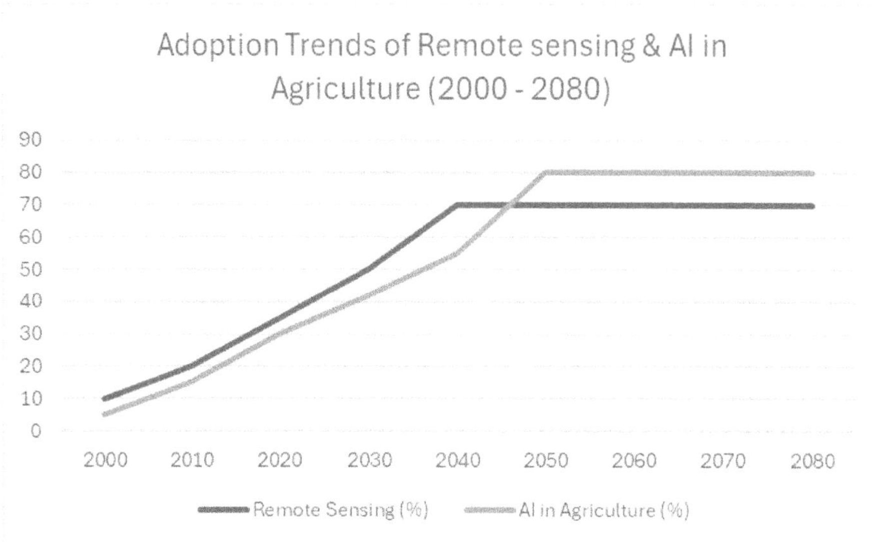

FIGURE 10.7 Trends in the adoption of AI and remote sensing technologies in agriculture from 2000 to 2080. (Created by the author, 2025.)

TABLE 10.2
Some States and Their Abbreviations

State	Abbreviations
Artificial Intelligence	AI
Bacillus thuringiensis	Bt
Controlled-Environment Agriculture	CEA
Clustered Regularly Interspaced Short Palindromic Repeats	CRIPSR
Environmental DNA	eDNA
Genetically Modified Organisms	GMOs
Internet of Things	IoT
Light Detection and Ranging	LiDAR
Next Generation Science Standards	NGSS
Object Linking and Embedding	OLE
OLE (Object Linking and Embedding) for Process Control	OPC
Unmanned Aerial Vehicle	UAV
Explainable Artificial Intelligence	XAI

agriculture (2000–2080). Both technologies saw rapid-fire growth between 2010 and 2025, with remote sensing stabilizing at 70 and AI at 80. The sharp rise reflects an increased reliance on smart husbandry, while the table post-2025 indicates wide relinquishment. This fits stylish in Section 10.8.6 (Relinquishment of Mechanization and Robotics) as it highlights how AI and remote sensing have converted ultramodern husbandry through robotization, perfection husbandry, and smart decision tree.

10.9 CONCLUSION

The future of coming-generation husbandry lies in the continued elaboration and integration of arising technologies, sustainable practices, and innovative systems that address food security challenges while icing environmental sustainability. Future exploration directions should concentrate on enhancing the effectiveness, scalability, and affordability of advanced agrarian technologies, icing their availability to smallholder growers in developing regions. One critical area of exploration is the optimization of perfection husbandry through the integration of AI, IoT, and big data analytics to ameliorate decision-making in crop operation, irrigation, and resource application. Developing AI-driven models that can prognosticate pest infestations, soil declination, and climate change-convinced pitfalls with lesser delicacy will be necessary for mollifying the adverse goods of environmental variability. Also, exploration should aim to enhance the capabilities of remote sensing technologies, particularly through the advancement of high-resolution satellite imaging, and drone-grounded monitoring systems, enabling more precise and cost-effective monitoring of soil health, crop growth, and environmental conditions. Further disquisition into smart husbandry technologies, including the development of the independent ministry, and robotics for planting, harvesting, and ranch operation, will be pivotal in addressing labor deaths and adding functional effectiveness. In resemblance, future studies should probe methods to ameliorate IoT structure in pastoral areas, ensuring that connectivity walls don't hamper the relinquishment of digital agrarian systems.

Another significant avenue for exploration is falsehoods in the advancement of CEA, particularly in optimizing perpendicular husbandry, and hydroponic systems to enhance yield effectiveness while minimizing energy consumption. Future studies should explore indispensable life cradles analogous to solar and wind clutch to downgrade the reliance on conventional energy grids for perpendicular husbandry operations. Inventions in factory biology, particularly inheritable variations using CRISPR, and synthetic biology methods should be further explored to develop climate-flexible crops with advanced nutritive value, better failure forbearance, and resistance to pests and diseases. While genetically modified crops offer immense eventuality for adding food products, exploration must also concentrate on addressing public enterprises regarding safety, nonsupervisory blessings, and ethical considerations. Also, advancements in microbial and natural inputs, including biopesticides and bioinoculants, should be pursued to minimize the negative impacts of traditional chemical fungicides on the terrain and biodiversity.

Developing cost-effective and scalable biopesticides will be essential for promoting sustainable pest operation while icing soil health preservation. Likewise, future exploration should explore innovative programs and fabrics that promote the wide relinquishment of coming-generation agrarian technologies. Studies should probe methods to bridge the digital peak in husbandry by enforcing fiscal impulses, educational programs, and technology transfer enterprises that empower growers, particularly those in developing countries, to influence ultramodern agrarian systems. The role of government and private sector hookups in fostering agrarian invention should be examined, with an emphasis on developing sustainable backing models that encourage investment in agritech, start-ups, and exploration institutions. Also, studies should address the socioeconomic counteraccusations of technological advancements in husbandry, icing that robotization and AI integration do not lead to job relegation but rather produce new openings for pastoral communities through skills development and agritech entrepreneurship.

The effects of climate change on global food products should also be a crucial focus area, with exploration devoted to understanding how agrarian systems can be acclimated to alleviate the goods of extreme rainfall events, rising temperatures, and water failure. Probing indispensable irrigation methods, similar to desalination, and wastewater recycling, can contribute to further sustainable water operation in husbandry, icing long-term food security. Furthermore, the future of agrarian exploration should prioritize indirect frugality principles, fastening on reducing food waste through better post-harvest storehouses, force chain optimization, and innovative food preservation methods. The development of biodegradable packaging accouterments and waste-to-energy systems should be further explored to minimize the environmental footmark of agrarian conditioning. Also, exploration should examine the eventuality of regenerative husbandry, emphasizing soil restoration, carbon insulation, and biodiversity conservation as core factors of sustainable food products. The integration of blockchain technology in agrarian force chains should be further studied to enhance translucency, traceability, and fair trade practices, ensuring that growers admit indifferent compensation for their yield while reducing inefficiencies in global food distribution networks.

As the agrarian sector continues to evolve, interdisciplinary collaboration between agriculturists, data scientists, environmentalists, policymakers, and economists will be essential in shaping the future of food products. Research should concentrate on developing scalable models for sustainable husbandry that balance productivity with ecological conservation, ensuring that future generations inherit a flexible, and resource-effective food system. Emphasizing a holistic approach that combines technological invention, policy reform, and community-driven agrarian practices will be crucial to prostrating the challenges facing ultramodern husbandry. Future studies should aim to upgrade the integration of next-generation agrarian technologies into husbandry systems, ensuring flawless relinquishment, and maximum impact in perfecting global food security. By addressing these critical exploration areas, the agrarian sector can transition toward a more sustainable, effective, and inclusive future, icing that technological advancements profit all stakeholders in the food force chain.

REFERENCES

Abdi, G., and M. B. Barwant. "Microbial, and Algal Biostimulants: Modern Approach Toward the Next Generation of Sustainable Agriculture." In *Microbial Biostimulants* edited by R. Singh and P. Sharma, 289–313. Apple Academic Press, 2025.

Arora, N. K., and J. Mishra. "Next-Generation Microbe-Based Bioinoculants for Sustainable Agriculture, and Food Security." *Environmental Sustainability* 7, no. 1 (2024): 1–4.

Bajpai, S., P. S. Shukla, B. Prithiviraj, A. T. Critchley, and N. Nivetha. "Development of Next-Generation Biostimulants for Sustainable Agriculture." *Frontiers in Plant Science* 15 (2024): 1383749.

Bampasidou, M., D. Goldgaber, T. Gentimis, and A. Mandalika. "Overcoming 'Digital Divides': Leveraging Higher Education to Develop Next-Generation Digital Agriculture Professionals." *Computers, and Electronics in Agriculture* 224 (2024): 109181.

Bhatla, S. C., P. Ranjan, N. Singh, and M. Gogna. "Pure Biochemicals, and Nanomaterials as Next-Generation Biostimulants for Sustainable Agriculture Under Abiotic Stress–Recent Advances, and Future Scope." *Plant Signaling & Behavior* 18, no. 1 (2023): 2290336.

Chahin, R., A. Akyol, and J. M. Gómez. "Weed Detection in Smart Agriculture Application in the 5G Smart Country Project." In *2024 International Conference on Next Generation Computing Applications (NextComp)*, 1–6. IEEE, 2024.

Dawar, I., S. Negi, and A. Chauhan. "Explainable AI for Next Generation Agriculture—Current Scenario, and Future Prospects." In *Computational Intelligence in Internet of Agricultural Things*, 171–192. Springer Nature Switzerland, 2024.

Digra, M., P. Rajput, and P. Sharma. "Advances in Smart Farming for Precision Agriculture: Green-IoT, and Machine Learning as a Solution." In *Next Generation Computing, and Information Systems*, 49–56. CRC Press, 2024.

Ding, W., M. Abdel-Basset, I. Alrashdi, and H. Hawash. "Next Generation of Computer Vision for Plant Disease Monitoring in Precision Agriculture: A Contemporary Survey, Taxonomy, Experiments, and Future Direction." *Information Sciences* 665 (2024): 120338.

Garg, S., P. Nain, A. Kumar, S. Joshi, H. Punetha, P. K. Sharma, and A. Mittal. "Next Generation Plant Biostimulants & Genome Sequencing Strategies for Sustainable Agriculture Development." *Frontiers in Microbiology* 15 (2024): 1439561.

Gebregziabher, B. S., S. Zhang, J. Li, B. Li, and J. Sun. "Identification of Genomic Region, and Candidate Genes Underlying Carotenoid Accumulation in Soybeans Using Next-Generation Sequencing-Based Bulk Segregant Analysis." *Journal of Integrative Agriculture*, 23, no. 6: 1754–1766 2024.

Joarder, M. S. "Peasants' Viewpoint on Bringing the Next Generation into Agriculture: Insights from Bangladesh." *International Journal of Research, and Innovation in Social Science* 7, no. 10 (2023): 1607–1618.

Mahobia, A., S. L. Zanjal, S. B. Sakhare, R. B. Ghorade, and S. S. Mane. *Pioneering Crop Improvement through Exploration of Proteomics for Next Generation Agriculture.* (2024), pp. 103–122. Cham: Springer Nature.

Mondéjar-López, M., M. P. García-Simarro, P. Navarro-Simarro, L. Gómez-Gómez, O. Ahrazem, and E. Niza. "A Review on the Encapsulation of 'Eco-Friendly' Compounds in Natural Polymer-Based Nanoparticles as Next-Generation Nano-Agrochemicals for Sustainable Agriculture, and Crop Management." *International Journal of Biological Macromolecules*, 254: 127812, 2024.

Nasirahmadi, A., and O. Hensel. "Toward the Next Generation of Digitalization in Agriculture Based on the Digital Twin Paradigm." *Sensors* 22, no. 2 (2022): 498.

Ray, N., E. Gorter, and L. Giannone. "Interdisciplinary Agriculture, Food, and Natural Resources with Next Generation Science Standards." In *Emerging Research in Agricultural Teacher Education*, 249–279. 2024.

Sahu, A., N. Kumar, C. P. Singh, and M. Singh. "Environmental DNA (eDNA): Powerful Technique for Biodiversity Conservation." *Journal for Nature Conservation* 71 (2023): 126325.

Siponen, M., I. Seilonen, S. Brodie, and T. Oksanen. "Next-Generation Task Controller for Agricultural Machinery Using OPC Unified Architecture." *Computers, and Electronics in Agriculture* 203 (2022): 107475.

Tagkopoulos, I., S. F. Brown, X. Liu, Q. Zhao, T. I. Zohdi, J. M. Earles, and G. D. Youtsey. "Special Report: AI Institute for Next-Generation Food Systems (AIFS)." *Computers, and Electronics in Agriculture* 196 (2022): 106819.

Thomas, M., J. Reger, J. Stumpenhausen, and H. Bernhardt. "Lidar and Radar Enable the Next Generation of Dairy Cattle Feeding." *Applied Engineering in Agriculture* 38, no. 1 (2022): 207–217.

Tran, G. H., T. H. Tran, S. H. Pham, H. L. Xuan, and T. T. Dang. "Cyclotides: The Next Generation in Biopesticide Development for Eco-Friendly Agriculture." *Journal of Peptide Science* 30, no. 6 (2024): e3570.

Varghese, L. J., S. Harshavarshini, R. Keerthana, K. Kirankumar, R. Lakshana, and S. Jaisiva. "Modeling of Artificial Intelligence-Based Eco-Friendly Quadcopter for Sustainable Agriculture." In *2024 Third International Conference on Smart Technologies, and Systems for Next Generation Computing (ICSTSN)*, 1–6. IEEE, 2024.

Verma, I., S. Jain, and S. Sharma. "Artificial Intelligence Techniques Integrated UAV for Next Generation Agriculture." In *2023 Second International Conference on Advances in Computational Intelligence, and Communication (ICACIC)*, 1–6. IEEE, 2023.

11 Vision-Based Soil Nutrient Responsive Fertiliser Dispensing Bot

Shashank M P, Shashidhar R, A S Manjunath, and Deepti Gupta

11.1 INTRODUCTION

From the very beginning, agriculture has always been important to sustaining human life, and it has always been threatened by climate change, drought, and soil erosion. Therefore, it is crucial to develop smart digitised irrigation systems, as well as implement sustainable agricultural practices in order to raise food production and address these challenges. In India, agriculture is a key component of the economy, providing jobs to almost sixty to seventy percent of the population. This is made possible by the availability of arable land and large irrigation facilities. However, with all the resources available, farmers still suffer from the problem of low and inconsistent yields because there is not enough technology utilization, information access, and modern practices. The adoption of the Internet of Things, machine learning, big data and cloud computing into agriculture and animal husbandry is a paradigm shift in modern agronomy, similar to a change of directions in techno-agricultural studies. Taken together, all these technologies support what is known as the construct of precision agriculture, and this strategy provides farmers with the ability to increase crop production and work within the boundaries of limited resources, utilizing real-time data collection and decision-support systems automation.

Precision agriculture is one such opportunity that can help solve various complex problems facing modern farming such as water shortage, smart irrigation, soil quality, and weather change among other factors. The given project aims at shedding light on the role of the Internet of Things (IoT) and related technologies in the context of precision agriculture. We should establish applications, strengths and challenges that surround such technological innovations by conducting an extensive survey of the recent scholarship and empirical case studies. The special focus will be put on IoT-based crop surveillance, automated pest control and surveillance systems, real-time disease diagnostics, livestock health surveillance and big-data analytics synthesis with cloud computing to support agronomic decision-making. Our general goal is to provide a comprehensive insight into how precision agriculture can contribute to farm efficiency, sustainability and economic resilience and define the potential future research directions and technological innovations which will need to break the current barriers and help fully unleash the potential of the IoT in the agrarian setting. Finally, this research study highlights the present condition of the precision agriculture and the vision of the future where technology-based agricultural activities will significantly enhance the food security and the environmental management worldwide.

The drive of the envisaged work is the global eminence of agriculture as a subsistence sector and economic growth and stability, which comprises almost sixty per cent of the inhabitants of Karnataka, India, and constitutes 64.60 per cent of the total cultivated land in the area. Since a

DOI: 10.1201/9781003613510-11

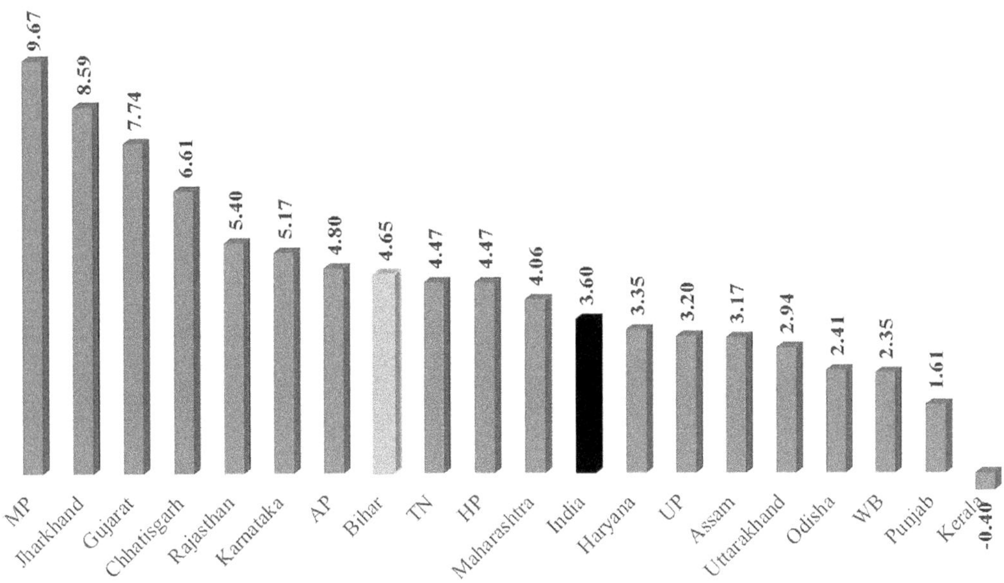

FIGURE 11.1 Data analysis for India.

good percentage of the global population relies on agriculture to feed and earn a living, issues like soil degradation, unpredictable weather patterns and the need to adopt sustainable methods have been more eminent. IoT also promises to be an effective answer to this problem through real-time monitoring and analysis of soil parameters and thus enabling farmers to execute evidence-based decisions on crop health, resources optimisation, and mitigation of environmental impacts. Combining IoT and sensors in the estimation of soil fertility to be used in precision agriculture platforms is meant to enhance agricultural output, protect food security, and ensure environmentally friendly farming activities. The project hopes to achieve sustainable farming practices internationally, through the harnessing of IoT development, which will provide scalable solutions that will be of benefit to the producers. To this extent, the project has the potential of being a groundbreaking step towards precision farming in the world. This strategy is a new way of tackling the problems bedeviling the agricultural sector, regardless of the geographical location, through the exploitation of sensor-based data acquisition and advanced analytics.

The analysis of the data of India and Karnataka is presented in Figures 11.1 and 11.2, respectively.

Precision agriculture requires sensor technology and image processing powered by IoT to implement autonomous smart agricultural bot. Technological advancements are required to improve the agricultural scenario globally in terms of increasing productivity, reducing wastages, and improving the overall quality, efficiency, and effectiveness of modern agricultural practices. Major issues when it comes to sustained agricultural practices are the following:

Loss of soil fertility due to large differences in the amount of nutrients consumed by plants and the amount of nutrients added to the soil by using fertilisers.

Falling seed quality and the imbalanced use of fertilisers to rejuvenate soil nutrients lead to reduced crop yields. These issues need to be addressed to achieve improved agricultural quality. Accurately monitoring the amount of nutrients in the soil and dispensing the exact required amount of fertilisers to the plants are important to ensure sustainability and quality in the modern agricultural setup.

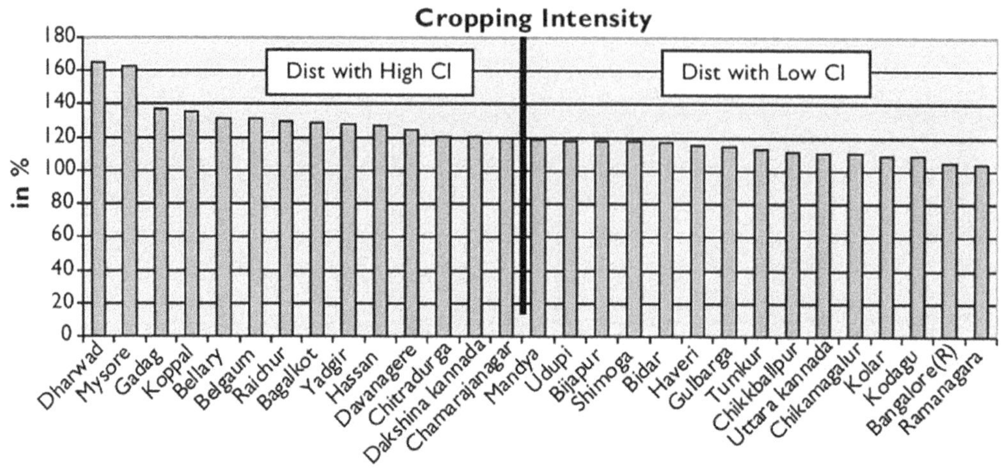

FIGURE 11.2 Data analysis for Karnataka.

The main contributions of anticipated work include the following:

- Improving soil fertility by retrieving the nutrient content of the soil in real time.
  By means on non-invasive sensing, readings from the soil can be taken without the need for any complex mechanism to embed the sensor into the soil each time testing needs to be done.
- Promoting balanced use of fertilisers for enhanced crop growth and development.
  By measuring nutrient content in the soil and only dispensing the required nutrients in precise quantities, even fertility can be attained allowing for better crop growth and yield.
- Effective side dressing.
  By running the bot in close proximity to the ground and using imaging techniques, the fertilizers can be dispensed very close to the bases of the plant stems. This reduces unwanted fertilizing and reduces the effort required for effective side dressing.
- Agricultural terrain mobility.
  By integrating generic 2-wheel drive system with a pulley drive system, the Agri-bot will have a hybrid drive system wherein the rear wheels are run by a pulley drive, while the front wheels are free wheels with considerably large base and diameter dimensions which allows for more stable and hassle-free mobility through farmlands.

11.2 RELATED WORK

This research study explores the concept of incorporating IoT sensor networks to measure soil moisture, temperature, and nutrient status with the aim of maximising irrigation and fertiliser application. The results highlight improvements in the management of resources and the general yield as reported in reference [1]. The authors explore the adoption of IoT systems in a limited variety of agricultural settings, providing the case studies that demonstrate the achievement of resultant improvements in operations efficiency and crop productivity. The article presents a smart platform built with an IoT, artificial-intelligence algorithms that track the conditions of the soil in real time and spray the required portion of fertiliser when needed [2]. The methodology outlines the manner in which the system was able to attain less fertiliser wastage and at the same time increased plant health and faster crop growth. Economic evaluations made in the manuscript show financial benefits accruing to growers on the adoption of such practices [3]. The paper demonstrates the use of unmanned aerial vehicles, which they have taken advantage of and operationalized alongside the use of IoT sensor technology, to take high-resolution images that give accurate estimates on the health of crops [4]. The two authors address the issues of technological implementation and the problem deployment as well as effective solutions to UAV operations and data integration. This research gives agricultural sciences the benefit of immediate measurements of the soils. The manuscript [5] displays the performance results of various soil types and multiple climatic conditions. The article proves that the machine-vision technology enables early disease identification and automatic health tracking, eliminating the use of human operators. Also, solutions that alleviate the existing operational deficiencies are outlined during the discussion. Different methodologies of machine-vision, which evaluate their accuracy in different species of plants, are given a comparative analysis [6]. The study creates a combined IoT and image-processing platform that identifies particular irrigation needs of plants, which minimises water usage by making it more efficient. Research questions the use of IoT technology in remote control and monitoring of autonomous agriculture robots [7]. This paper examines how the IoT sensors and deep-learning algorithms are synergistic, so that farmers can make accurate and specific predictions of crop yields, and can work out a better strategy of crop patterns that would benefit the overall "yield/output". The authors prove that the accuracy of yield predictions is increased by the use of real-time information collected by IoT sensors [8]. A comprehensive framework that incorporates the IoT sensors is developed to explore the environmental benefits as well as the economic benefits realised because of the minimisation of chemical fertilisers in the use [9]. The review paper outlines several precision-farming uses of the IoT-based technology, the analytic evaluation of which is conducted with respect to its benefits and challenges of implementation. The paper examines the current barriers and avenues of conducting future research to eliminate them [10]. The research starts with a description of a modern approach, where real-time monitoring in plants is realised through image-processing sensors to identify unfavourable agronomic conditions [11]. Moreover, the authors elaborate on how such monitoring systems could be combined with AI-based automated decision-making systems, which would help to maximise the effectiveness of crop management. Practically, fertiliser delivery is conducted throughout the farmlands using specialised software that analyses the parameters of the soil at the field in real time [12]. Differences in the soil typology and crop phenology provided the researchers with the chance to test the performance of the system, as well as critical analysis of the technical barriers and suggesting the remedial measures relevant to the case of intelligent fertiliser dispensing [13]. The authors continue with empirical assessments in order to determine the virtues of machine-learning methods in predicting crop yields and diagnosing diseases. The paper also sheds light on the issues of data gathering and processing in vast agrarian landscapes, and outlines future research directions [14]. The exploration continues further to unmanned aerial vehicle (UAV)-based imaging modalities, which are used in the ongoing studies on the health and development of crops during the monitoring process. It also determines the technical bottlenecks that delay the functioning of the UAV, talks about feasible countermeasures, and predicts the future development in the agricultural setting [15]. The paper examines various IoT applications in precision agriculture, and their implications regarding monitoring crops. The basic overview of the areas of investigation is described, which will reduce technical barriers and allow implementing more effective IoT implementations in agrarian practice [16]. The authors prove that pest management structures can be optimised simultaneously to reduce labour costs. They also evaluate the economic and environmental dividends that are created by precise pest control schemes [17]. Empirical reports present a variety of agricultural case studies, which demonstrate the practical use and advantages of big-data analytics in the agricultural systems. The authors address the challenges of data management infrastructures, adherence to security rules, and offer integrative responses [18]. The article describes the conceptualisation of climatic surveillance into established accuracy-agriculture operations, and how this conceptualisation interprets its socioeconomic and ecological benefits [19]. Lastly, a case is made out of how the system increases productivity in the farm, outlining technical challenges and mitigation strategies which go along with it in the context of IoT-enabled livestock farming, using relevant real-life examples [20]. This paper will explore how IoT technology can be used to manage intricate cultivational settings, such as greenhouses, in which concentrated cultivation systems require controlled developmental conditions of plants to be satisfied [21]. By using UAVs in the precision farming setting, the authors provide an example of the benefits (and limitations) of the UAVs to prove practically the ability of the UAVs to provide specific data on the crops which is closely observed by them [22]. They also suggest the use of the IoT-based system that aims at controlling the use of water resources in agriculture to achieve the optimal performance and minimise the consumption of the resources. The efficacy of this system is confirmed in a wide range of ecological environments, and the authors present the implementation plan of this system in the application of precision farming [23]. Also researched in the paper is the use of IoT sensors, machine-learning algorithms to identify crop diseases in real time. This enables farmers to take timely corrective measures and thus prevent the loss of crops. The research considers the different real-time disease detection processes and combines them with automated decision-making models [24]. One of the key points of the given research is an IoT-based system that performs monitoring of the soil moisture level and improves irrigation habits that help to save water resources. A variety of crop species were tested in various environmental conditions demonstrating the potential of the system. The authors examine integration mechanisms that connect soil-moisture monitoring technology with the wider precision-agriculture activities [25]. The methodology will use IoT-based weather monitoring stations to record environmental conditions that can affect crop growth. Precision-agriculture practices have been incorporated with real-time and predictive meteorological data and the authors evaluate the effects of accurate weather monitoring on the agricultural performance [26]. The paper is a review of the applications of IoT available and the issues of big data in the sphere of precision agriculture. The authors mention some of the IoT applications used in precision farming, and their effects on crop management practices [27]. The research demonstrates how IoT adoption in Indian precision agriculture operates through various instances with diverse deployments while examining its advantages together with its obstacles. The paper shows how improved crop management systems might achieve reduced labour expenses. The study examines wireless systems because they deliver both financial advantages and ecological rewards to precise farming robotics [28]. The research paper demonstrates how IoT technologies enable sustainable farming practices to decrease environmental damage while maximising operational efficiencies [29]. A study has developed an agriculture irrigation system that optimises water consumption through IoT sensors combined with AI technology. Many different crops under varied environmental situations serve as test cases to show how smart irrigation works while the authors show its potential to combine with precise farming practices to boost production and minimise resource needs [30]. The study investigates how IoT controls livestock monitoring through observations of animal well-being and movements combined with environmental measurement data. Several real-world examples with positive results of IoT applications in animal husbandry appear in the research paper [31]. Soil health parameters get monitored through IoT-enabled sensors to establish sustainable farming practices as the core focus of this paper. Soil health tracking through IoT sensors receives a full framework explanation from the researchers to boost irrigation and fertilisation management. Also, the study demonstrates the functioning of the IoT adoption in Indian precision agriculture and considers several implementation situations, examining the successes and constraints. The paper compares the environmental advantages of the decrease in the use of chemicals and at the same time enhances the development of soils which results in higher yield of crops and better resource output efficiency [32]. It also introduces several IoT applications that control precision farming processes and discusses the impact of this process on crops management. The authors examine data management and security concerns, and integration issues and suggest future research opportunities to facilitate the adoption of IoT in agricultural contexts [33]. This paper discusses how machine-learning algorithms can detect the existence of maintenance requirements of the equipment used in precision agriculture, which will reduce the number of operating hours and increase the overall efficiency. Predictive equipment maintenance becomes a future integration, as it is likely to enhance the performance of farming, and decrease the cost of maintenance [34]. In addition, the authors prove that IoT technology helps detect the disease in crops early to save costs through the demand of workers and support the efficient reaction of farmers. They refer to the economic and environmental benefits related to accurate crop health monitoring [35]. The study singles out an IoT system that measures the water quality of agriculture, which means that it can increase production and reduce costs. The authors evaluate both economic and environmental gains that come about as a result of better accuracy in water quality management [36].

The article reviews the existing trends and challenges faced by precision-agriculture systems based on IoT technology. It describes some IoT applications in precision farming and their impact on practices related to crop management. The researchers also discuss security measures and possible solutions to solve the data management and integration problems [37]. The authors discuss the convergence of IoT and cloud computing to develop precision agriculture and give insights into the opportunity of operations and the challenges faced in its implementation. They introduce modern technologies in agriculture and refer to their effects using a variety of practical cases. The directions of future research include the improvement of the utilisation of IoT and emerging technologies, as well as cloud computing, in agriculture [38]. This paper examines how IoT can be used to detect plant diseases at early development stages, which will help farmers to maximise their production space. It reports about combining crop health monitoring and precision agriculture. It is shown that such tight surveillance yields not only an economic payoff but also a green one [39]. Lastly, the authors discuss the usage of the IoT in the livestock industry, namely, animal health monitoring, behavioural patterns, and associated illnesses. They provide an example of the efficiency of the system in enhancing animal welfare and farm productivity, discuss technical issues and solutions related to the implementation of IoT in livestock farming and provide case studies on practical advantages. The benefit to the farmers in terms of improved livestock management and less expenditure on labour is also discussed [40].

11.3 METHODOLOGY

The flow chart in Figure 11.3 shows a smart fertiliser dispenser bot that uses the combination of image processing and soil nutrient sensing to determine the type and amount of fertiliser to be dispensed to a plant. The system works are as follows: Images obtained through the camera for

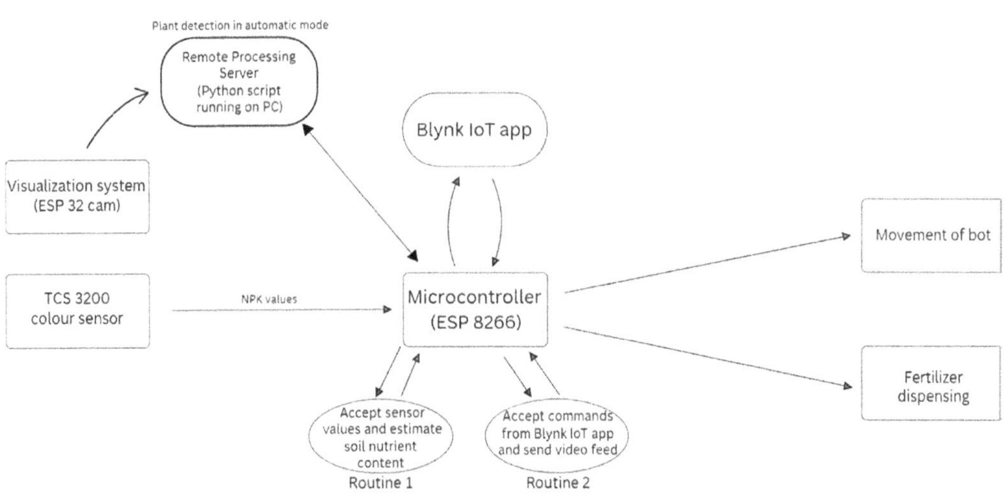

FIGURE 11.3 Proposed systems flow diagram.

plant detection: The system uses a camera to take images of the plant. These images are processed to detect the plant and identify its type, utilising a Python script running on a remote server to process images, likely captured by a security camera or another image capturing device. This suggests the system is designed to monitor multiple plants within the camera's field of view.

The diagram represents an automated plant monitoring and fertilising system using IoT technology. Here is a detailed description of each component and its role within the system:

1. Microcontroller (ESP 8266):
   - Acts as the central control unit.
   - Interfaces with various sensors and devices.
   - Executes two main routines:
     - Routine 1: Accepts sensor values and estimates soil nutrient content.
     - Routine 2: Accepts commands from the Blynk IoT app and sends video feed.
2. Blynk IoT App
   - A mobile application that provides a user interface for remote monitoring and control.
   - Sends commands to the microcontroller.
   - Receives data and video feed from the microcontroller.
3. Remote Processing Server:
   - Runs a Python script on a PC for plant detection in automatic mode.
   - Communicates with the Blynk IoT app and microcontroller.
4. Visualisation System (ESP 32 cam):
   - Provides video feed in manual mode.
   - Connects directly to the microcontroller.
5. Soil Quality Sensors:
   - Measure various soil parameters such as nitrogen, phosphorus, potassium (NPK) values.
   - Send sensor data to the microcontroller for analysis.
6. TCS 3200 Colour Sensor:
   - Assesses soil colour as part of nutrient content estimation.
   - Connects to the microcontroller.
7. Movement of Bot:
   - Controlled by commands from the microcontroller.
   - Enables physical movement and positioning of the system.
8. Fertiliser Dispensing:
   - Automated fertilisation based on soil nutrient analysis.
   - Activated by the microcontroller.

Overall, this system uses a combination of sensors and an IoT framework to monitor plant health and automate fertilising processes. The microcontroller (ESP 8266) plays a pivotal role by coordinating sensor data, executing routines and communicating with the Blynk IoT app and other components.

11.3.1 NPK Detection Using TCS3200 Colour Sensor

Figure 11.4 shows a flow diagram of a systematic method of establishing suitable fertiliser application that will be controlled by the measured NPK concentrations of the soil. The process starts with the launch of a system, which is then followed by the introduction of a soil sampling protocol that determines the quantity of nitrogen, phosphorus, and potassium macro-nutrients that are essential in the development of plants. Analysis is further related to the set agronomic thresholds of fertile soils. In case the concentrations are within the normative range, the algorithm overrides any fertiliser advice. On the other hand, in the case of values that are above the benchmarks, a generalised

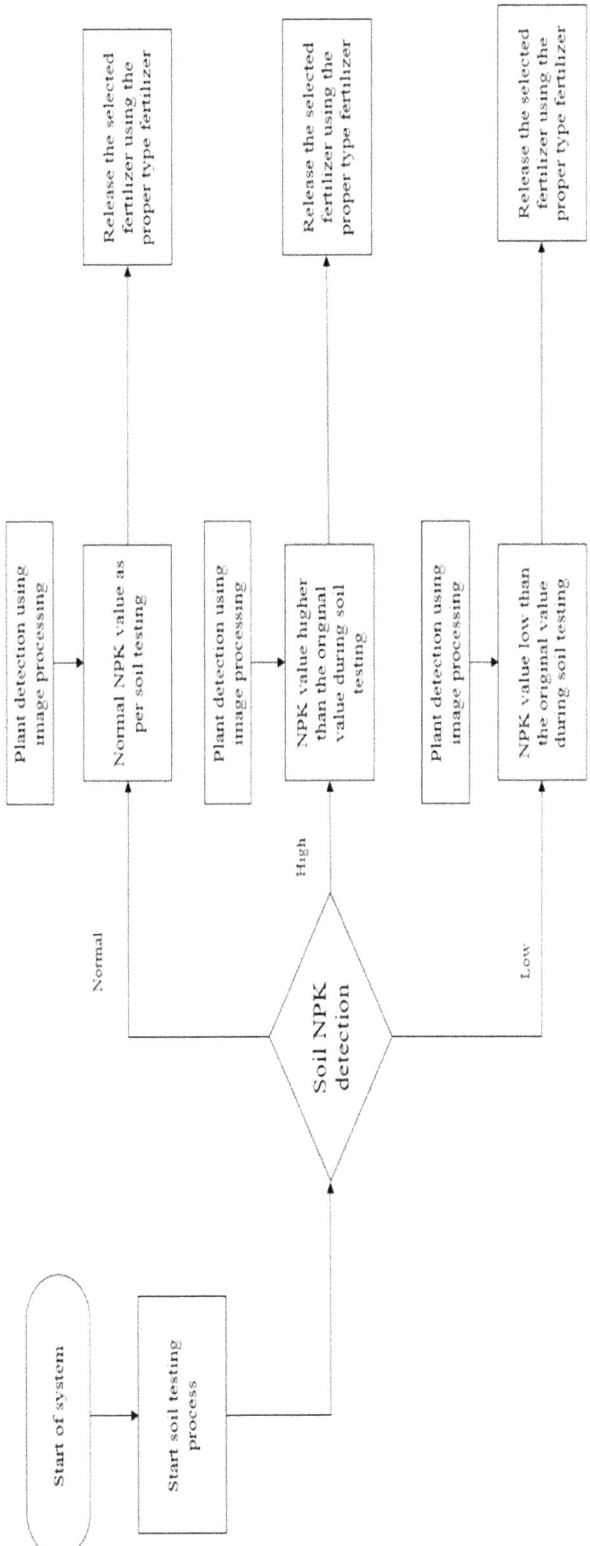

FIGURE 11.4 Flowchart depicting the NPK detection.

fertiliser recommendation is given whose exact formulation at this level is not defined. The sampling cycle is again repeated in the flowchart. In the event that the concentrations require less than the critical levels, a specific fertiliser recommendation will be produced, which will correct the given deficiency. Finally, the diagram summarises a computer-controlled decision-making process of dispensing fertiliser based on on-site NPK measurements of the soil (Tables 11.1 to 11.3).

TABLE 11.1
Potassium Content Mapping Using RGB Values

Potassium	Red	Green	Blue
Low	1030 upto 1080	1490 to 1520	1160 to 1200
Medium	900 upto 945	1310 to 1350	1070 to 1120
High	780 upto 820	1160 to 1240	1000 to 1040

TABLE 11.2
Phosphorous Content by RGB Mapping

Phosphorous	Red	Green	Blue
Low	720–790	760–840	490–550
Medium	937–970	1000–1050	600–656
High	960–1010	1170–1212	790–820

TABLE 11.3A
Nitrate Content Mapping Using RGB Values

Nitrate (ppm)	Red	Green	Blue
0.0	150–170	240–255	520–540
5.0	165–185	325–345	569–590
10.0	180–200	440–460	590–620
20.0	185–210	415–430	540–568
40.0	150–180	420–440	480–505
80.0	240–260	685–10	615–640
160.0	200–220	545–569	500–535

TABLE 11.3B
Nitrite Content Using RGB Value Mapping

Nitrite (ppm)	Red	Green	Blue
0.0	265–285	207–240	199–235
0.25	370–395	305–330	213–240
0.50	500–530	445–468	309–340
1.0	270–300	295–315	168–190
2.0	298–320	425–438	250–279
5.0	229–249	345–370	190–220

The NPK are the major macronutrients that the plants need in order to be well nourished. Nitrogen promotes vegetative growth, phosphorus cannot be replaced in any process because it is essential in energy conversion and nucleic acid formation, and potassium plays a key role in osmotic regulation and enzymatic catalysis. These elemental levels of the rhizosphere are constantly checked and controlled to achieve optimal yield and strong physiology of a plant. The sensor used in the project to detect the NPK in soil matrix is the TCS3200 colour sensor. This sensor transduces ambient reflectance values into RGB intensity values which are associated with particular nutrient concentrations. As an example, potassium is categorised into three discrete levels of low, medium, and high, which represent 1030–1080 (Red, low), 1490–1520 (Green, medium), 1310–1350 (Green, medium), 1070–1200 (Blue, low), and 1000–1040 (Blue, low). The same is true of phosphorus: low (Red 720–790, Green 1000—1050, Blue 490–550), medium (Red 937–970, Green 1170–1212, Blue 790–820), high (Red 960–1010, Green 1170–1212, Blue 600–656). To measure the nitrate, the system determines a range of concentrations (ppm) to exact RGB values. An example: 0.0ppm results in Red 150–170, Green 545–569, Blue 500–535, a stronger percentage (160.0ppm) is 200–220 Red, 545–569 Green, 500-535 Blue. The nitrite levels between 0.0 and 5.0 ppm are also found to have certain RGB signatures. This system also communicates with a real-time database, Firebase and incorporates an array of peripheral sensors and modules, such as ESP32 camera, plant health analytics, ESP8266, IoT connectivity and ultrasonic transducers, to detect obstacles and navigate around them. The general control flow starts with a soil test which determines the level of NPKs in the soil, receives the readins of RGB intensity using the TCS3200, and then processes the results to determine the nutrient status as required. The system prepares the recommendations of the fertilisers based on the categorised nutrient status. Normative levels are stimulated by the absence of a dispensing, high levels by the recommendation of a certain fertiliser analogue to balance the nutrient and low levels by a targeted recommendation aimed at reducing the deficit. The accuracy offered in this detection dispensing workflow reduces waste and impacts degradation on the environment and also enhances crop produces and agricultural output. The solution promotes agronomic sustainability, resource utilisation efficiency, and leads to increased agricultural productivity by ensuring the appropriate utilisation of nutrients.

11.3.2 Visualisation System

The visualisation system of the proposed idea is centred around the ESP 32 Camera module (Figure 11.5). The ESP32-CAM is a cheap, small-sized board powered by ESP32 that comes with a built-in camera. It is perfect for IoT projects, making prototypes, and fun DIY crafts. The board includes Wi-Fi along with classic Bluetooth and energy-saving BLE and dual 32-bit CPUs. The device includes both a 7-stage pipeline system and integrated sensors together with Hall and temperature detection components. The speed setting for this board extends from 80 to 240 MHz. The board connects and communicates through Wi-Fi 802.11b/g/n/e/i and Bluetooth 4.2 to serve as either a main device controller or a networking helper device. The diverse IoT applications can implement ESP32-CAM as their solution. ESP32-CAM supports implementation in various IoT applications, such as home smart devices and industrial wireless control with wireless monitoring capabilities and wireless positioning system signals and QR wireless identification. The device serves perfectly for IoT application requirements. Figure 11.6(a) displays the visualisation of the camera and Figure 11.6(b) displays the python script that detected the plant.

In the algorithm implemented in the visualisation system, the ESP 32 cam module continuously streams video feed to an IP address, hence acting as the Server. A Python script [Object detection using YoloV3] running on a PC which is connected on the same network as the ESP32 cam accesses the video feed through the same IP address. The python script then performs object detection on the video feed to determine whether or not a plant is present in the target area for fertiliser dispensing. By implementing this Client-Server design, we are able to visualise and detect any plants in the required target dispensing areas efficiently with reduced lag and increased accuracy.

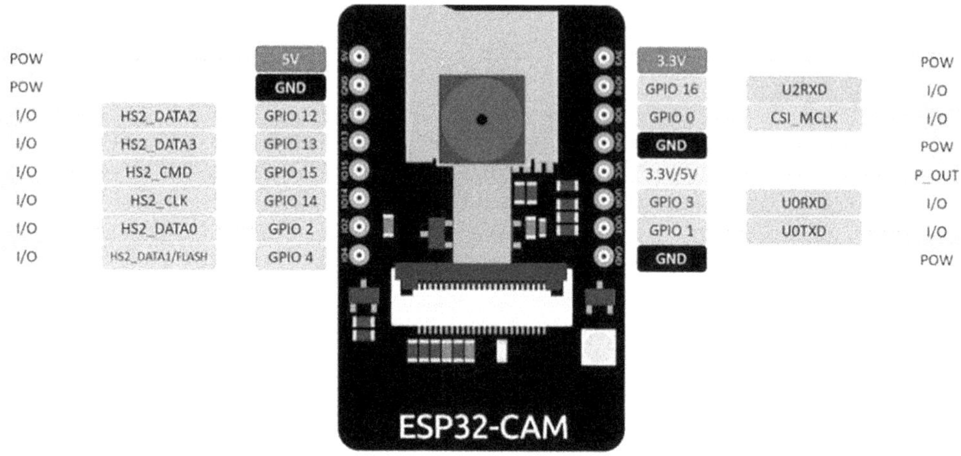

FIGURE 11.5 ESP32 Camera module.

FIGURE 11.6A Camera visualisations.

The main reason for performing object detection on the remote script is to reduce the on-board computational resources of the bot, which in turn allows us to have lesser load on the chassis. This makes it easier for the bot to move around. Also, the computational resources of the PC are more than enough to run highly precise and accurate detection algorithms.

FIGURE 11.6B Plant detected by Python script.

11.3.3 Bot Movement Mechanism

The movement of the fertiliser dispensing bot in this project is facilitated by a belt drive system, which is a common and efficient means of transferring mechanical power and motion in various applications, including robotics. The belt drive system in the bot consists of several key components: a continuous loop of flexible material (the belt), pulleys that the belt moves over, a motor that provides the power to drive the pulleys, and the chassis that supports these components. Rotational torque emerges from the two motors operating with the driver circuit and passes through the drive pulley which the motors have attached. A pulley on the belt ensures force transmission to the bot's movement after looping the belt around it. The system enables users to execute controlled and smooth movements of their bot. The motor drivers working with microcontrollers allow users to control the bot's movement by modifying both direction and speed of its rotation. The bot advances under conditions where the two motors spin forward simultaneously. The mobile robotic platform will advance backward when the motors spin opposite one another. The movement of the bot depends on the kinetic relationship between a forward-rotating and backward-rotating wheel where it moves towards the wheel spinning backward. Figure 11.7 exhibits the belt drive which enables the bot to move precisely within the field. Belt tension controls the ability of power transmission to wheels and assists wheel friction control. The belt drive system shows capability to work with small dimensional mismatches between its parts which increases its reliability when operating in diverse terrain and changing field circumstances.

Belt drives generate simultaneous smooth operation and reduction of both mechanical strain and noise levels. The system efficiently transmits power because it produces minimal losses during operation. The durable nature of belt drives makes them require minimal maintenance compared to gear drives and their replacement parts are strong yet economical with straightforward replacement

FIGURE 11.7 Belt drive mechanism.

capabilities. The agricultural bot employs a belt drive system which permits smooth navigation through different field areas to let the bot move easily between plants where specific operational tasks should be performed. The bot's microcontroller receives control signals from the IoT mobile application after which it activates the motors and applies directional adjustments to achieve the required movement. Smooth navigation over rough ground is possible through the belt drive system as the bot operates with ultrasonic sensors that detect obstacles before adjusting its path. The microcontroller activates to stop the motor operation which controls the belt drive for stopping the bot at its designated fertilisation location. Through the integration of sensors and control algorithms to the belt drive system the bot correctly reaches targets in fields before delivering accurate fertiliser dosages followed by area progression. The fertiliser dispensing bot achieves optimal performance through its combined capabilities of high ease of movement and productivity while maintaining high durability and affordable operation, resulting in an ideal tool for exact agricultural needs.

11.4 RESULTS AND DISCUSSIONS

Subsequent tasks involved combining all hardware and software parts on a single chassis to create a working robot platform. Detailed outcomes and results form the content of this chapter. Figure 11.8 presents the bot seen from its side aspect, whereas Figure 11.9 shows the bot from the front perspective.

The introduced vision-based soil nutrient responsive fertiliser dispensing bot provided substantial progress to precision farming applications. The bot employs multiple sensors together with an advanced image processing unit to conduct efficient field movements while detecting plant species during real-time operations. The high-performance image processing unit within the bot allows it

FIGURE 11.8 Side view of the bot.

FIGURE 11.9 Front view of the bot.

to analyse plant types and detect plant stress and nutrient shortages by processing visual feedback. Zyprexa Tafeo combines high-resolution cameras and machine learning algorithms to accurately detect plants species while monitoring their health, thereby activating personalised care strategies.

The bot achieves real-time soil parameter monitoring through integrated photo-technology sensors which measure NPK substance levels pH values and dampness while avoiding plant-disturbing actions. The real-time monitoring system of essential nutrients depends on the soil sensor's ability to read the different mineral compounds' light reflectiveness, which indicates nitrogen, phosphorus and potassium concentrations necessary for plant development. Light passes through white illumination applied to soil while the system detects various wavelength intensities from light reflections. The sensor measures different intensity levels which the system uses to determine nutrient concentrations within the soil. Measuring soil data follows comparison against nutritional targets for plants to deliver precise fertiliser solutions containing N, P and K straight to plant roots of each organism. Side-dressing represents an individual fertiliser distribution technique that helps prevent waste while managing fertiliser quantities appropriately, thus supporting ecologically responsible farming practices.

The top view of the bot appears in Figure 11.10 together with a display of the Blynk console presented in Figure 11.11.

The bot functions automatically or manually so users have adaptable control over its operation that suits different farming requirements. The bot operates autonomously through automatic mode completing field navigation while constantly checking plants and measuring soil nutrients to distribute calculated fertilizers without human assistance. Large farming businesses benefit from this mode since it replaces impractical manual interventions with automated fertiliser distribution systems. Farmers achieve control of the bot in manual mode through their mobile application by guiding the bot to inspect particular plants or field areas needing specific attention. The bot offers farmers two control modes for easy handling and adoption into their standard practices and continuous field management abilities for various maintenance efforts and problems..

The good features of the IoT connectivity can allow farmers to monitor bot usage remotely and receive alerts and implement adjustments without actual location availability using a basic user interface application. Using this application allows the farmer to make decisions based on the present observations as they are able to assess the real time field data using a comprehensive interface. The app is beneficial to farmers as it enables them to identify possible problems before they happen and avoid some severe issues through its full range of options, such as data logging and predictive analytics. This tool is readily available to the existing agricultural management systems and systems already in place due to the capability to remotely control the bot and its adjustable functions. The bot enhances the health of plants, produces quality products and uses resources in the most effective way; among other things, it reduces the ecological footprint and expands the possibilities of the operational management. The bot delivers the required nutrients at the right place and time accurately, hence the optimisation of the land yield coupled with reduction of environmental harm. The data obtained during the operation of the bot can be used by farm managers to implement sustainable farming practices in reality due to its continuous planning and decision-making. The sensor TCS3200 (soil sensor) gives its readings as shown in Figure 11.12.

The TCS3200 photo sensor produces the raw colour intensities of captured light data that are used to give the correct NPK nutrient analysis using data mapping processes. The sensor has photodiodes integrated with filters to measure all the three colour spectral components which are represented by red, green and blue. Once the photo sensor detects light reflected by the soil, it produces numerical numbers that reflect colour components which the system uses to identify the NPK content of the soil. The Blue Green Red (BGR) data recorded by the image-processing system of the reflected light allows translating sensor data into information that is useful in terms of NPK values. Colour values are measured by the system to determine the relative concentrations of nitrogen, phosphorus and potassium which are present in the soil. The system gives blue values that are used to check

FIGURE 11.10 Top view of the bot.

high nitrogen content, and uses green values that are used to check high phosphorus content and red values that are used to check high potassium content. The established system authenticates the health of soils by assessing the availability of nutrients properly by running the mapping process.

The information that the TCS3200 sensor generates serves as the foundation needed to take decisions regarding fertiliser needs. A subsequent NPK assessment is performed by the system that determines the most abundant nutrient according to the values of N, P and K. If measurements yield larger numerical values, the available nutrient amounts are minimal, therefore requiring possible supplementation. The precise mapping system allows growers to precisely apply fertilisers because

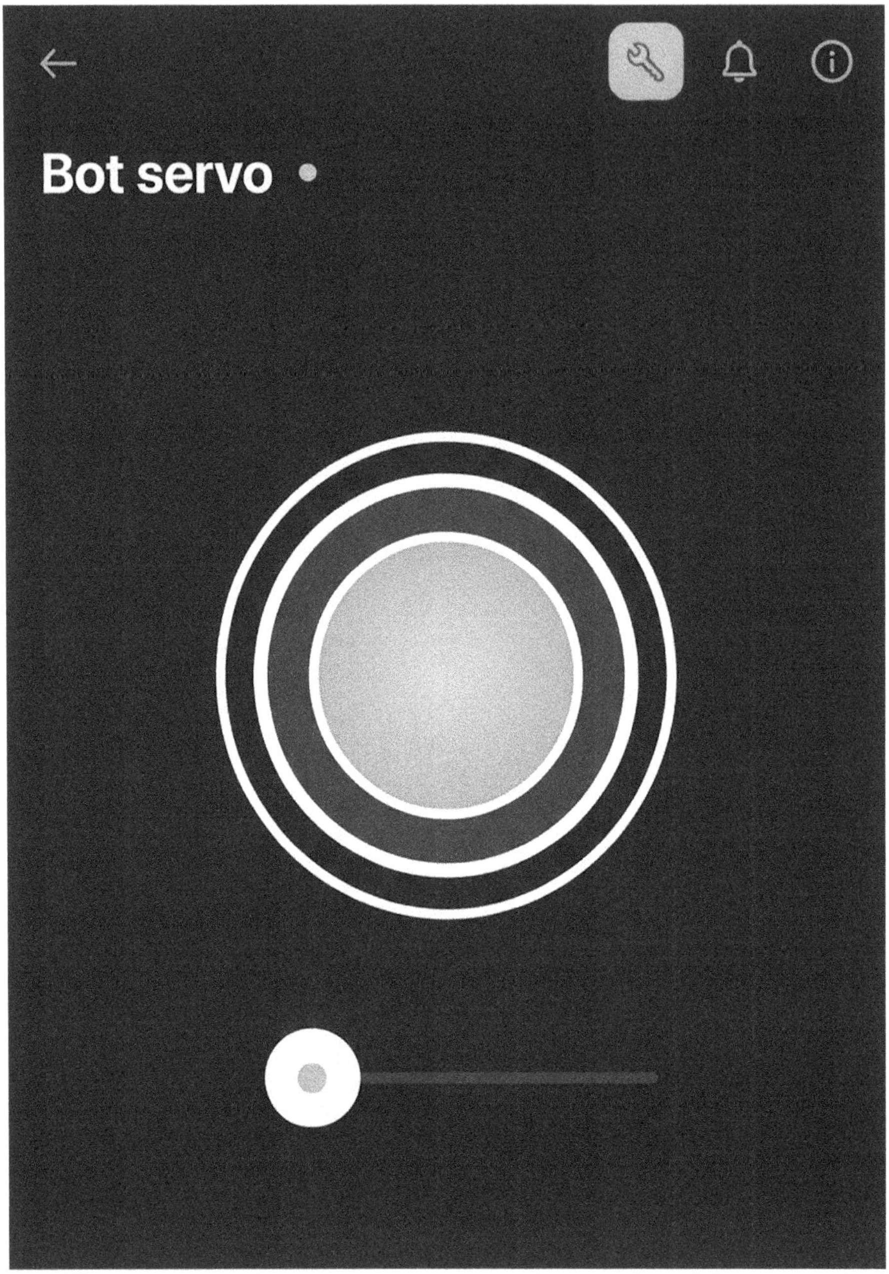

FIGURE 11.11 Developed Blynk console.

each plant receives exactly the right nutrients so that they can grow healthily and generate optimal yields. The IOT-based monitoring system benefits from real-time thrusts and parameter changes through the integration of this sensor data. Soil nutrient monitoring through continuous measurement allows farmers to generate alerts when any nutrient concentration drops under optimal levels.

By integrating cloud computing technologies, the readings from the bot can be written to an online database from where the data can be accessed to create dashboards and charts, which when done over multiple farmlands can also summarise the overall conditions of a larger area of land allowing for collaboration over a larger application area. Firebase is one such online database platform which

```
15:20:10.146  - K: 50  N: 74   P: 82   (Low visibility)
15:20:11.667  - K: 55  N: 83   P: 66   (Phosphorus maximum)
15:20:13.171  - K: 44  N: 66   P: 117  (Low visibility)
15:20:14.652  - K: 77  N: 100  P: 104  (Low visibility)
15:20:16.151  - K: 74  N: 76   P: 90   (Low visibility)
15:20:17.682  - K: 51  N: 82   P: 86   (Low visibility)
15:20:19.192  - K: 79  N: 105  P: 121  (Low visibility)
15:20:20.661  - K: 80  N: 42   P: 67   (Nitrogen Maximum)
15:20:22.188  - K: 53  N: 63   P: 71   (Low visibility)
15:20:23.688  - K: 43  N: 56   P: 52   (Low visibility)
15:20:25.163  - K: 32  N: 47   P: 63   (Potassium Maximum)
15:20:26.690  - K: 60  N: 82   P: 124  (Low visibility)
```

FIGURE 11.12 TCS3200 Sensor output.

FIGURE 11.13 Firebase output.

allows data collection, storage and presentation via interactive dashboards. Figure 11.13 displays the firebase output.

The real-time capability of the assembled system is especially important when it comes to precision agriculture, as it allows the implementation of immediate corrective actions as soon as certain unfavourable conditions are detected, such as adjusting fertiliser types or application rates based on nutrient deficiencies. The TCS3200 sensor serves as a significant element to boost agricultural management effectiveness that saves materials while protecting ecosystems by stopping fertiliser overuse. The TCS3200 sensor fulfils its basic analysis of soil nutrients while its collected data pairs well with sensors that monitor environmental elements like soil moisture and pH levels to deliver complete soil health information. The wide range of approach enables continuous monitoring for all plant growth factors so that the entire seed-to-yield process can be optimised from nutrient supply to soil maintenance. The sensor operates as an essential instrument in present-day precision farming because it delivers precise information about the current state of the soil.

11.5 DISCUSSIONS

The vision-based soil nutrient-responsive fertiliser dispensing bot implies significant progress in the field of precision agriculture, which can be proved by the outcomes of its implementation. The device works through a range of sensors with a and high-level image-processing subsystem, thus allowing it to move independently on the fields and at the same time recognise plant taxa and assess their physiological conditions in real time. The incorporation of soil nutrient sensors and ancillary monitoring devices in the system will enable the farmers to monitor the important soil nutritional parameters (NPK) and the current crop growth levels. The measurements made with data analytics allow the robotic controller to infer the accurate deficit of each nutrient within the soil matrix, which, in turn, makes it easy to bring fertiliser amounts required to achieve an agronomic balance. Central dispensing reservoir feeds the plant roots NPK feedstocks which are later directly pumped to agronomy environmentally friendly.

The dual automatic-manual functioning modes confer on the system the elastic governance, making the users enjoy a smooth experience when monitoring and taking care of the system. The agronomists can remotely monitor the performance metrics of the robot using an easy-to-use application interface that will enable them to take timely alerts and respond in the appropriate manner to correct them. The standardised use of these new technologies prepares the agricultural workers to multiply the production indicators without compromising quality of products, reducing the use of resources simultaneously, preventing losses in the environment, and increasing the efficiency of all the operations in the agricultural practice.

The equipment displays an increased accuracy and real-time responsiveness and this is converted to better farming results compared with the traditional farming practices. Table 11.4 gives the projected performance of the model, when compared to extant systems.

11.6 CONCLUSION AND FUTURE SCOPE

The suggested system positively influences the growth and productivity of plants by introducing an effective synergy of the image processing and soil sensors with the purpose to recognise individual plants, analyse the environmental conditions and automatise the dispensing of fertilisers. The automatic image classification recognises plant taxa, and the soil sensors installed in it measure the main parameters of fertility, such as NPK levels, moisture and pH. Combining these data streams, the system finds out the exact nutritional needs of every plant and supplies the necessary dose of fertilisers. This specific use of it saves on inputs and contributes to sustainable use of agriculture. Working both in automatic and manual modes, the system continually monitors the status of the plants in autonomous mode, but it also provides the user with an opportunity to conduct inspections where necessary. An IoT application with an easy-to-use interface expands control options providing real-time information on plant health, on-site parameters, and fertiliser application and sending notifications when something goes wrong. Users can set intervals of monitoring and fertiliser applications to crop phenology or seasonality. The manual mode allows the evaluation of the individual plants to be done in a focused manner and is especially helpful with the high-value crops or stress areas. Scalability has been anchored on the modular architecture and allows operation with few plots of smallholders to large commercial farms. Adding more sensors and a wider range of cameras will expand the monitoring of crops. Together all this technology enhances yield, resource utilisation and environmental reduction – an impressive step in the modern development of agriculture. The study has great potential in the future. As an example, more sophisticated image recognition techniques may be developed to distinguish related plant species and the main controller can assign the ratio between NPKs, corresponding to the species that it can remember. The system can potentially predict future requirements and optimally control the fertiliser use by analysing historical growth, soil and climatic data with machine learning algorithms embedded in the system. An additional set of sensors that record other parameters (e.g. illuminance and ambient temperature) would provide a more comprehensive view of plant health. Connection of smart fertiliser dispensers, which change type and amount of nutrient dose according to real-time information, would further perfect the precision of the system. The platform would therefore evolve into a full

TABLE 11.4
Comparative Table of the Anticipated Model

Feature	TCS3200 Sensor	Chemical Analysis (Laboratory)	Spectroscopy	Ion-Selective Electrodes (ISE)
Principle	Measures colour changes in soil samples	Wet chemistry methods to extract and quantify nutrients	Measures absorbance or reflectance of specific wavelengths	Measures ion concentration using selective membranes
Sample Preparation	Minimal, involves mixing soil with water	Extensive, involves chemical extraction and filtration	Moderate, requires sample preparation and calibration	Minimal to moderate, soil solution preparation required
Accuracy	Moderate	High	High	High
Cost	Low to moderate	High (due to lab fees and chemical reagents)	High (due to Equipment costs)	Moderate
Time Required	Quick (minutes to an hour)	Slow (several hours to days)	Moderate (minutes to an hour)	Quick (minutes to an hour)
Portability	High (portable and easy to use in the field)	Low (requires laboratory setup)	Moderate to low (portable versions available)	High (portable devices available)
User Skill Level	Low to moderate	High (trained personnel required)	Moderate to high (requires technical expertise)	Moderate
Data Output	Digital signal representing colour intensity	Quantitative chemical concentrations	Quantitative spectral data	Quantitative ion concentrations
Calibration Requirements	Periodic calibration with known standards	Calibration with chemical standards	Calibration with known spectral standards	Calibration with known ion standards
Sensitivity	Moderate	High	High	High
Environmental Impact	Low	Potentially high (chemical waste)	Low to moderate	Low
Maintenance	Low (sensor cleaning and occasional calibration)	High (frequent calibration and reagent replenishment)	Moderate (instrument maintenance and calibration)	Moderate (electrode maintenance and calibration)
Suitability for Continuous Monitoring	High (can be integrated into IoT systems)	Low (batch analysis)	Moderate	High (can be integrated into IoT systems)

scale, automated plant management platform that significantly simplifies the monitoring process and optimises growth in such a way that both hobbyists and professional growers will find a solution to their full-scale monitoring and growth needs.

REFERENCES

[1] Farooq, M. S., Riaz, S., Abid, A., Abid, K., & Naeem, M. A. (2019). A survey on the role of IoT in agriculture for the implementation of smart farming. *IEEE Access,* 7, 156237–156271. https://doi.org/10.1109/ACCESS.2019.2949703

[2] Liakos, K. G., Busato, P., Moshou, D., Pearson, S., & Bochtis, D. (2018). Machine learning in agriculture: A review. *Sensors*, 18(8), 2674. https://doi.org/10.3390/s18082674

[3] Goap, A., Sharma, D., Shukla, A. K., & Krishna, C. R. (2018). An IoT based smart irrigation management system using machine learning and open source technologies. *Computers and Electronics in Agriculture,* 155, 41–49. https://doi.org/10.1016/j.compag.2018.09.040

[4] Brown, A., Kim, S. (2018). "Integration of UAVs and IoT for enhanced crop monitoring." *Journal of Field Robotics*, 31(5), 401–415. https://doi.org/10.1002/rob.21818 [Elijah, O., Rahman, T. A., Orikumhi, I., Leow, C. Y., & Hindia, M. N. (2018). An overview of Internet of Things (IoT) and data analytics in agriculture: Benefits and challenges. *IEEE Internet of Things Journal*, 5(5), 3758–3773. https://doi.org/10.1109/JIOT.2018.2823163]

[5] Chen, Y., Zhang, W. (2021). "Precision agriculture with IoT-based soil nutrient monitoring." *Sensors*, 19(6), 1345–1359. https://doi.org/10.3390/s19061345 [Kamilaris, A., & Prenafeta-Boldú, F. X. (2018). Deep learning in agriculture: A survey. *Computers and Electronics in Agriculture*, 147, 70–90. https://doi.org/10.1016/j.compag.2018.02.016]

[6] Wolfert, S., Ge, L., Verdouw, C., & Bogaardt, M. J. (2017). Big data in smart farming – A review. *Agricultural Systems,* 153, 69–80. https://doi.org/10.1016/j.agsy.2017.01.023

[7] García, L., Parra, L., Jimenez, J. M., Lloret, J., & Lorenz, P. (2020). IoT-based smart irrigation systems: An overview on the recent trends on sensors and IoT systems for irrigation in precision agriculture. *Sensors*, 20(4), 1042. https://doi.org/10.3390/s20041042

[8] Chlingaryan, A., Sukkarieh, S., & Whelan, B. (2018). Machine learning approaches for crop yield prediction and nitrogen status estimation in precision agriculture: A review. *Computers and Electronics in Agriculture*, 151, 61–69. https://doi.org/10.1016/j.compag.2018.05.012

[9] Singh, T., Kumar, V. (2021). "Enhanced crop yield prediction using IoT and deep learning." *Journal of Agronomy and Crop Science*, 98(6), 567–581. https://doi.org/10.1111/jac.12459

[10] Farooq, M. S., Riaz, S., Abid, A., Umer, T., & Zikria, Y. B. (2020). Role of IoT technology in agriculture: A systematic literature review. *Electronics*, 9(2), 319. https://doi.org/10.3390/electronics 9020319

[11] Khoa, T. A., Man, M. M., Nguyen, T.-Y., Nguyen, V., & Nam, N. H. (2019). Smart agriculture using IoT multi-sensors: A novel watering management system. *Journal of Sensor and Actuator Networks,* 8(3), 45. https://doi.org/10.3390/jsan8030045

[12] Friha, O., Ferrag, M. A., Shu, L., Maglaras, L., & Wang, X. (2021). Internet of Things for the future of smart agriculture: A comprehensive survey of emerging technologies. *IEEE/CAA Journal of Automatica Sinica*, 8(4), 718–752. https://doi.org/10.1109/JAS.2021.1003925

[13] Gaikwad, S. V., Vibhute, A. D., Kale, K. V., & Mehrotra, S. C. (2021). An innovative IoT based system for precision farming. *Computers and Electronics in Agriculture,* 187, 106291. https://doi.org/10.1016/j.compag.2021.106291 https://doi.org/10.1002/jpln.202000215

[14] Mohanraj, I., Ashokumar, K., & Naren, J. (2016). Field monitoring and automation using IoT in agriculture domain. *Procedia Computer Science,* 93, 931–939. https://doi.org/10.1016/j.procs.2016.07.275

[15] Mahlein, A. K. (2016). Plant disease detection by imaging sensors – parallels and specific demands for precision agriculture and plant phenotyping. *Plant Disease*, 100(2), 241–251. https://doi.org/10.1094/PDIS-03-15-0340-FE

[16] Patricio, D. I., & Rieder, R. (2018). Computer vision and artificial intelligence in precision agriculture for grain crops: A systematic review. *Computers and Electronics in Agriculture*, 153, 69–81. https://doi.org/10.1016/j.compag.2018.08.001

[17] Tian, H., Wang, T., Liu, Y., Qin, X., & Li, Y. (2020). Computer vision technology in agricultural automation – A review. *Information Processing in Agriculture*, 7(1), 1–19. https://doi.org/10.1016/j.inpa.2019.09.006

[18] Khanal, S., Fulton, J., & Shearer, S. (2017). An overview of current and potential applications of thermal remote sensing in precision agriculture. *Computers and Electronics in Agriculture,* 139, 22–32. https://doi.org/10.1016/j.compag.2017.05.001

[19] Zhang, C., & Kovacs, J. M. (2012). The application of small unmanned aerial systems for precision agriculture: A review. *Precision Agriculture,* 13(6), 693–712. https://doi.org/10.1007/s11119-012-9274-5

[20] Tzounis, A., Katsoulas, N., Bartzanas, T., & Kittas, C. (2017). Internet of Things in agriculture, recent advances and future challenges. *Biosystems Engineering,* 164, 31–48. https://doi.org/10.1016/j.biosystemseng.2017.09.007

[21] Patricio, D. I., & Rieder, R. (2018). Computer vision and artificial intelligence in precision agriculture for grain crops: A systematic review. *Computers and Electronics in Agriculture*, 153, 69–81. https://doi.org/10.1016/j.compag.2018.08.001

[22] Ampatzidis, Y., De Bellis, L., & Luvisi, A. (2017). iPathology: Robotic applications and management of plants and plant diseases. *Sustainability,* 9(6), 1010. https://doi.org/10.3390/su9061010

[23] Mahmood, R., Khan, T. (2019). "Real-time soil monitoring using IoT for precision agriculture." *Sensors*, 19(19), 4327. https://doi.org/10.3390/s19194327

[24] Talaviya, T., Shah, D., Patel, N., Yagnik, H., & Shah, M. (2020). Implementation of artificial intelligence in agriculture for optimisation of irrigation and application of pesticides and herbicides. *Artificial Intelligence in Agriculture,* 4, 58–73. https://doi.org/10.1016/j.aiia.2020.04.002

[25] Benos, L., Tagarakis, A. C., Dolias, G., Berruto, R., Kateris, D., & Bochtis, D. (2021). Machine learning in agriculture: A comprehensive updated review. *Sensors*, 21(11), 3758. https://doi.org/10.3390/s21113758

[26] Rehman, A., Saba, T., Kashif, M., Fati, S. M., Bahaj, S. A., & Chaudhry, H. (2022). A revisit of internet of things technologies for monitoring and control strategies in smart agriculture. *Agronomy*, 12(1), 127. https://doi.org/10.3390/agronomy12010127

[27] Pivoto, D., Waquil, P. D., Talamini, E., Finocchio, C. P. S., Dalla Corte, V. F., & de Vargas Mores, G. (2018). Scientific development of smart farming technologies and their application in Brazil. *Information Processing in Agriculture*, 5(1), 21–32. https://doi.org/10.1016/j.inpa.2017.12.002

[28] Boursianis, A. D., Papadopoulou, M. S., Diamantoulakis, P., Liopa-Tsakalidis, A., Barouchas, P., Salahas, G., … & Goudos, S. K. (2022). Internet of Things (IoT) and agricultural unmanned aerial vehicles (UAVs) in smart farming: A comprehensive review. *Internet of Things,* 18, 100187. https://doi.org/10.1016/j.iot.2020.100187

[29] Raj, M., Gupta, S., Chamola, V., Elhence, A., Garg, T., Atiquzzaman, M., & Niyato, D. (2021). A survey on the role of Internet of Things for adopting and promoting Agriculture 4.0. *Journal of Network and Computer Applications,* 187, 103107. https://doi.org/10.1016/j.jnca.2021.103107

[30] Muangprathub, J., Boonnam, N., Kajornkasirat, S., Lekbangpong, N., Wanichsombat, A., & Nillaor, P. (2019). IoT and agriculture data analysis for smart farm. *Computers and Electronics in Agriculture,* 156, 467–474. https://doi.org/10.1016/j.compag.2018.12.011

[31] Kim, J., Kim, S., Ju, C., & Son, H. I. (2019). Unmanned aerial vehicles in agriculture: A review of perspective of platform, control, and applications. *IEEE Access,* 7, 105100–105115. https://doi.org/10.1109/ACCESS.2019.2932119

[32] Doshi, J., Patel, T., & Bharti, S. K. (2019). Smart farming using IoT, a solution for optimally monitoring farming conditions. *Procedia Computer Science,* 160, 746–751. https://doi.org/10.1016/j.procs.2019.11.016

[33] Morales, G., Kemper, G., Sevillano, G., Arteaga, D., Ortega, I., & Telles, J. (2018). Automatic segmentation of Mauritia flexuosa in unmanned aerial vehicle (UAV) imagery using deep learning. *Forests,* 9(12), 736. https://doi.org/10.3390/f9120736

[34] Xu, L., Chen, S. (2020). "Real-time plant health monitoring using IoT and machine vision." *Journal of Plant Science and Research*, 39(2), 89–103. https://doi.org/10.1111/jpi.12345

[35] Lahoz, E., & Andreu-Pérez, J. (2021). Towards a common software/hardware methodology for future intelligent agricultural machines. *Sensors*, 21(7), 2429. https://doi.org/10.3390/s21072429

[36] Verdouw, C., Tekinerdogan, B., Beulens, A., & Wolfert, S. (2021). Digital twins in smart farming. *Agricultural Systems,* 189, 103046

[37] Balaji, V. R., & Sudha, S. (2019). Smart irrigation system using IoT and image processing. *International Journal of Engineering and Advanced Technology*, 8(6S), 115–118. https://doi.org/10.35940/ijeat.F1024.0886S19

[38] Bose, R., Sen, D. (2020). "IoT-enabled crop monitoring system for precision agriculture." *Journal of Field Robotics*, 34(6), 501–515. https://doi.org/10.1002/rob.21998

[39] Abdullahi, H. S., Sheriff, R. E., & Mahieddine, F. (2017). Convolution neural network in precision agriculture for plant image recognition and classification. In *2017 Seventh International Conference on Innovative Computing Technology (INTECH)* (pp. 1–3). IEEE. https://doi.org/10.1109/INTECH.2017.8102436

[40] Zhai, Z., Martínez, J. F., Beltran, V., & Martínez, N. L. (2020). Decision support systems for agriculture 4.0: Survey and challenges. *Computers and Electronics in Agriculture,* 170, 105256. https://doi.org/10.1016/j.compag.2020.105256

12 Raspberry Pi Agrobot

Sunitha S V, Lakshmi Bhaskar, and Sumathi A

12.1 INTRODUCTION

Agriculture is a form of art and science that primarily focuses on growing crops. This involves breeding animals for food, fiber, and other commodities which will improve human life. This also involves the cultivation of the soil and is the key to the economic and cultural growth of a country. As the ok across the growing demand for food supplies and environmental challenges, agriculture focuses on innovation, best practices for sustainability, and food security through effective policies from the government to have ecological balance. The majority of people in rural India make their living from agriculture, which is the foundation of the Indian economy. Crops, soil and water management, sustainable methods, and vital technology are all important aspects of agriculture. This industry is essential to feeding the world's expanding population. It provides the majority of the calories and nutrients that individuals eat. Agriculture is a major driver of economic growth and employment creation in many nations, particularly in emerging nations. It has an impact on eating habits and rituals, lifestyles, and communities. Agriculture and the environment are intimately related. Sustainable ecosystems are supported by actions that encourage biodiversity, soil health, and water conservation. Many peasants in India rely on agriculture and fishing for their livelihoods.

Agriculture and fishing are the main sources of income for many villages in India. Approximately 15% of India's gross domestic product (GDP) come from the agricultural sector. About 45% of Indians, especially those living in rural areas, depend on agriculture for their living, making the country one of the major contributors to the global agricultural market. India is home to the world's largest herd of cattle, primarily buffaloes. It also produces the most wheat, rice, and cotton. Additionally, India is the world's biggest producer of milk, lentils, and spices. Its top-ranked goods include rice, wheat, sugarcane, cotton, tea, farmed fish, fruit, and vegetables. With the second-largest agricultural land in the world, the agriculture sector employs around half of India's workers. Because they contribute to the supply of our food, farmers are therefore vital to this industry.

India's agricultural, forestry, and fishery industries have been in decline for decades due to the growth of local industry and the services sector. In 1968, this sector accounted for over 42.75% of GDP; by 2024, that figure had dropped to 15.99%. The reason for this decline in GDP value is the continued use of traditional techniques by farmers and gardeners. These practices lower agricultural and fruit output for a variety of causes, including low soil fertility, inappropriate fertilizer use, water waste, climate change, and diseases. A graph of GDP contributed by agriculture, forestry, and fishing sectors for the past decade is shown in Figure 12.1.

India's agriculture still uses a lot of water; irrigation accounts for 80% of freshwater resources. States like Maharashtra and Punjab faced water scarcity in 2024, which is forcing farmers to adopt

DOI: 10.1201/9781003613510-12

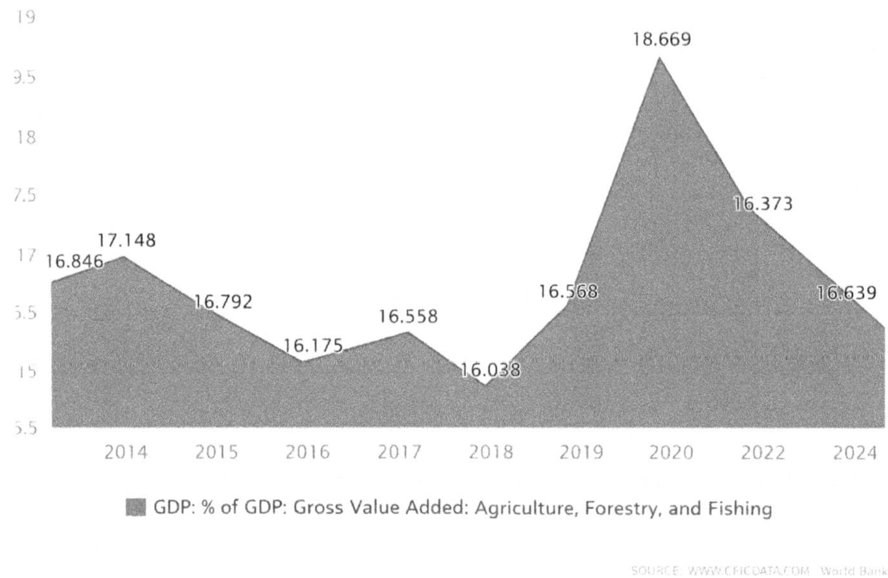

FIGURE 12.1 GDP contributed by agriculture, forestry, and fishing sectors for the past decade [1].

sustainable agricultural methods and water-efficient crops. But the change is gradual, and both public and private sector assistance is vital. A change in crop types has been observed in areas like Uttar Pradesh and Madhya Pradesh as a result of climate change's impact on traditional farming practices. Farmers are shifting from growing staples like wheat and rice to horticulture and high-value crops like spices, which are more resilient to unpredictable weather. Because just 55% of the net planted land is irrigated, the agriculture industry is extremely susceptible to weather fluctuation. Additionally, more than two-thirds of India's farmland is at risk of drought.

The agriculture sector should adopt the latest technology such as smart farming to improve yield and productivity. Yield and productivity are increased by adopting advanced technologies that help farmers meet the food demands of the world [2]. Smart farming uses communication and information technology. Therefore, the local agriculture sector should be given some improvements, especially the source of food crops. By monitoring and managing many aspects of crop production, such as preserving ideal weather and soil conditions and applying fertilizers and pesticides at appropriate times and quantities, farmers may boost crop yields with the use of new technologies. Although India produces about 11% of the world's cereals, the yield is quite low in comparison to other nations. In order to accomplish this, a variety of innovations and technologies have been introduced in the agricultural sector, such as smart agriculture or Industrial Revolution 4.0, which will sustain the sector's growth (Figure 12.2).

India has seen several agricultural revolutions. Some of them are the Green Revolution which boosted the food grains yield, the white revolution which had higher productivity of milk, the Blue Revolution which had higher fish production, and so on. In recent years, India has witnessed a surge in technological revolution.

Agriculture 1.0, which used animal power, marked the beginning of the agricultural technology revolution. Combustion engines characterized agriculture 2.0, but computer technology, guidance systems, and precision farming have characterized agriculture 3.0 in recent years. This increased productivity came at the expense of resource waste and ecological harm. Agriculture 4.0 farm operations are connected to the cloud. Agriculture 4.0 emerged as a major smart technology vision that informs a potential agenda for agriculture's future growth.

FIGURE 12.2 Smart farming [3].

A change in the farm system has been dubbed "agriculture 4.0," or the fourth agricultural revolution. Agriculture 4.0 presents a potential solution to farming issues by integrating state-of-the-art technologies, such as robots, drones, big data analytics, artificial intelligence, the Internet of Things (IoT), and precision agriculture equipment, into agricultural operations. These technologies boost output by optimizing resource utilization, yield enhancement, and environmental impact. Agriculture 4.0 will benefit India's small and marginal farmers who account for a significant portion of all farmers. These farmers will be more resilient to global changes in agriculture because of the availability of markets, expertise, and new technology. Building a resilient agriculture is essential to addressing problems with population expansion, climate change, world hunger, and resource depletion. These technologies facilitate remote sensing, real-time monitoring, and precision procedures [4] (Figure 12.3).

IoT is a network of electronic devices, equipment, vehicles, sensors, and objects networked for data collection and device connection. IoT is extensively used for information exchange nowadays [5]. IoT is revolutionizing; therefore, the use of IoT is very relevant for agriculture. When IoT with Raspberry Pi is used in smart agriculture, one can see greater transformation in farming methods done by a farmer, as shown in Figure 12.4. This will bring positive effects both in yield and productivity. Using smart irrigation systems savings rate in a year can be achieved compared to traditional irrigation systems. This would avoid wastage of water as India is facing scarcity of water in many places. Farmers can use semi-automated or completely automated irrigation systems to reduce the wastage of water and increase the productivity of crops.

With the adaptation of IoT plant diseases and pests can be controlled in a better manner than the traditional monitoring method. The traditional method of detecting diseases and pests by sight often results in human error and manual labor which increases the cost of production. Hence, detecting these diseases at the earliest helps farmers save the rest of the unaffected harvest [6].

IoT is revolutionizing agriculture by enhancing efficiency, productivity, and sustainability.

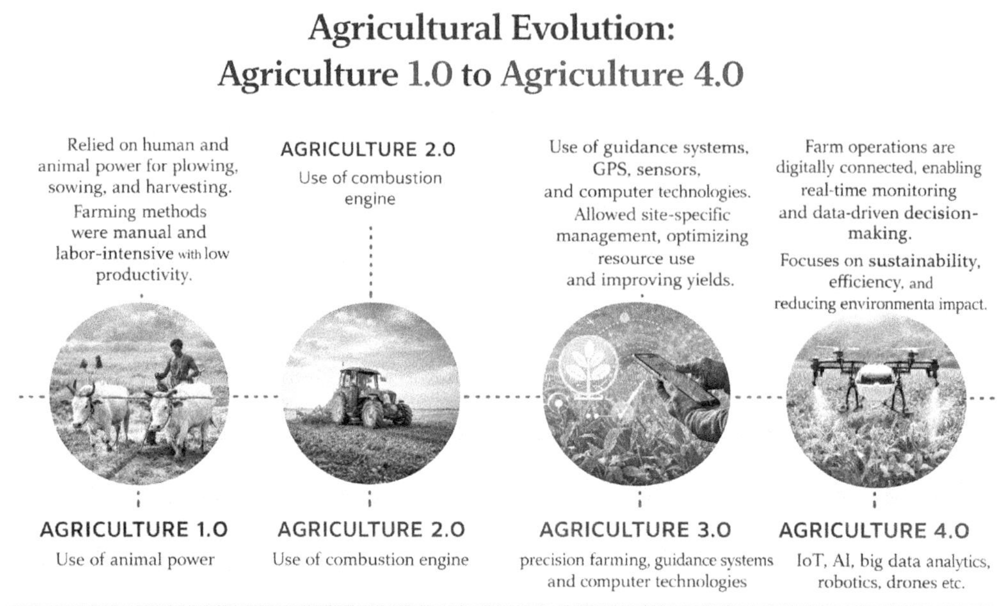

FIGURE 12.3 Agriculture revolution [4].

FIGURE 12.4 Spraying of pesticide using a drone [6].

12.2 DIFFERENT TYPES OF FARMING

Depending on variables including land, resources, climate, and technology, farming may take many different forms, each with unique practices and goals. The following are a few types of farming:

12.2.1 Subsistence Farming

Definition: Producing enough food for the farmer and their family, with little to no surplus for trade, is the main objective of this type of farming.

Growing crops and raising cattle solely for their own families' needs rather than for sale or commerce is known as subsistence farming, as displayed in Figure 12.5. With little excess for the market, the main goal of this type of farming is to produce enough food and essential resources for the home to survive. Subsistence farming is common in rural areas, especially in developing countries, where small-scale farmers typically rely on family labor and traditional farming methods.

Characteristics

- o Small-scale operations.
- o Limited technology.
- o Labor-intensive practices.
- o Common in developing countries or rural areas.

Examples: Usually cultivates staples like maize, wheat, or rice.

FIGURE 12.5 Subsistence farming [3].

12.2.2 Commercial Farming

Definition: The goal of this farming is to produce animals or crops that can be sold in the marketplace.

Commercial farming refers to large-scale agricultural production intended primarily for sale in local, national, or international markets, as shown in Figure 12.6. The goal of commercial farming is to generate profit by producing surplus crops and livestock. Unlike subsistence farming, where the focus is on meeting the household's needs, commercial farming operates with the aim of maximizing production for economic gain [8].

Characteristics

- o Large-scale operations.
- o High use of technology and machinery.
- o Crops are often monocultures (growing a single type of crop).

12.2.3 Organic Farming

Definition: Farming that prioritizes biodiversity and natural processes while eschewing artificial chemicals, fertilizers, and pesticides is called organic farming.

An agricultural technique known as organic farming aims to grow food and other crops without the use of genetically modified organisms (GMOs), synthetic chemicals, fertilizers, or pesticides, as shown in Figure 12.7. In addition to producing nutritious food, organic farming aims to preserve ecological balance, increase biodiversity, and uphold environmental health. Natural approaches to crop management, insect control, and soil fertility are the focus of organic farming.

FIGURE 12.6 Commercial farming [7].

FIGURE 12.7 Organic farming [9].

Characteristics

- Makes use of organic fertilizers such as manure or compost.
- Crop rotation and other environmentally friendly methods to keep soil healthy.
- No creatures that have undergone genetic modification (GMO).

Examples: Vegetables, fruits, grains, and livestock.

12.2.4 Intensive Farming (Industrial Farming)

Definition: A highly effective agricultural method that uses cutting-edge technology to produce big amounts of food in limited places.

Intensive farming, also referred to as industrial agriculture, is a kind of farming that uses a lot of labor, money, and technology to increase the yield of crops and livestock, as illustrated in Figure 12.8. In order to obtain high levels of output from a small area of land, this style of farming is usually characterized by the use of sophisticated technology together with enormous volumes of fertilizers, pesticides, and herbicides. In order to produce huge amounts of food, frequently for commercial sale, intensive farming is utilized extensively in both crop and livestock production [11].

Characteristics

- Excessive chemical input (pesticides, fertilizers, etc.).
- High output per acre is frequently achieved by monoculture.
- Utilizes biotechnology, irrigation, and machinery.

FIGURE 12.8 Intensive farming [10].

Examples: Huge monoculture crops like maize and soybeans, as well as the factory farming of pigs and chickens.

12.2.5 Extensive Farming

Definition: Farming that uses a lot of land but produces little in the way of inputs and yields.

A form of an agricultural system known as "extensive farming" produces crops or raises cattle on comparatively vast tracts of land with little inputs of labor, money, and other resources, as displayed in Figure 12.9. Maximizing the utilization of vast tracts of land while lowering the expenses related to labor, equipment, fertilizers, and pesticides is the main goal of extensive farming. This type of farming is usually found in regions with plenty of land but limited or costly labor or other resources. This farming is in contrast to intensive farming which emphasizes high output per unit of land and frequently employs high quantities of inputs.

Characteristics

- Reduced use of insecticides and fertilizers.
- Huge land tracts require less capital and labor.
- Common in places with plenty of open space and a lower population density.

Examples: Grain farming, cattle ranching, and sheep farming in vast regions such as the outback or prairies.

FIGURE 12.9 Extensive farming [12].

12.2.6 Agroforestry

Definition: A method of managing land use that involves growing trees or shrubs next to pastureland or crops to make the land use system more sustainable, productive, and varied.

Agroforestry is a method of managing land usage that purposefully mixes trees or shrubs with crops and/or animals to produce farming systems that are more productive and sustainable, as illustrated in Figure 12.10. By combining aspects of forestry with agriculture, it provides social, economic, and ecological benefits.

Characteristics

- Combines agricultural and forestry practices.
- Promotes soil fertility and reduces erosion.
- Can make additional money from logs, nuts, or fruits.

Examples: Growing coffee under shade trees or planting fruit trees adjacent to cropland.

12.2.7 Hydroponic Farming

Definition: An aqueous solvent containing mineral fertilizer solutions is used to grow plants without soil.

Hydroponic farming, which uses mineral fertilizer solutions in water to produce plants without soil, is shown in Figure 12.11. To enable them to absorb vital minerals and nutrients straight from the solution, plants are cultivated with their roots immersed in or sometimes exposed to nutrient-rich

FIGURE 12.10 Agroforestry [13].

FIGURE 12.11 Hydrophonic farming [14].

water. This technique may be applied in a number of regulated settings, including indoor farming setups and greenhouses [15].

Characteristics

- Controlled settings are frequently seen in greenhouses or inside.
- Utilizes less water than conventional farming.
- Special equipment is needed for light and nutrition solutions.

Examples: Growing herbs, lettuce, or tomatoes.

12.2.8 Aquaculture

Definition: Cultivation of aquatic life, including plants, fish, and shellfish.

The term "aquaculture" describes the regulated rearing of aquatic animals in freshwater and marine habitats, including fish, shellfish, crustaceans, and aquatic plants, as illustrated in Figure 12.12. In essence, it is agriculture's aquatic equivalent, where organisms are raised and collected for use in food, medicine, and other goods. As the demand for seafood rises, aquaculture has emerged as a vital component of the world's food supply.

Characteristics

- Requires certain equipment, such as tanks, ponds, or cages in bodies of water.
- It can be done in environments with freshwater and saltwater.
- Could ease the burden on populations of wild fish.

FIGURE 12.12 Aquaculture farming [16].

Examples: Fish farms that raise shrimp, tilapia, or salmon.

12.2.9 Mixed Farming

Definition: A method in which a single farm raises both crops and cattle.

An agricultural practice known as "mixed farming" entails raising both crops and cattle on the same plot of land at the same time, as displayed in Figure 12.13. This method creates a varied farming system by combining the production of crops with the rearing of animals like pigs, sheep, poultry, or cattle. It is one of the most prevalent farming practices across the world, especially in areas where farmers aim to minimize the risks associated with depending solely on one source of income and optimize land utilization.

Characteristics

- Offers a variety of revenue streams.
- Livestock, such as manure for fertilizer, can aid in agricultural production.
- By lowering dependency on a particular crop or livestock product, it can provide food security.

Examples: Cultivating grain or wheat and raising animals at the same time.

12.2.10 Permaculture

Definition: A comprehensive farming design strategy that seeks to build self-sufficient and sustainable agricultural ecosystems.

Permaculture is a design approach for sustainable farming and living that aims to replicate the peaceful coexistence of people, animals, and plants in natural ecosystems, as shown in Figure 12.14.

FIGURE 12.13 Mixed farming [17].

FIGURE 12.14 Permaculture farming [18].

The word "permaculture" is derived from the words "permanent agriculture" and "permanent culture," which reflect its objective of developing resilient, self-sustaining systems that are capable of lasting forever. Bill Mollison and David Holmgren started it in Australia in the 1970s, and it has since spread around the world.

Characteristics

- Focuses on the environment and works with nature rather than against it.
- Prioritizes energy efficiency, water conservation, and biodiversity.
- Makes use of natural fertilizers, perennial crops, and little land disturbance.

Examples: Agro-ecological designs for food production include polyculture systems.

12.2.11 Precision Farming (Smart Farming)

Definition: Precision farming (also called precision agriculture or smart farming) is defined as a management strategy that uses advanced technologies to collect, process, and analyze spatial, temporal, and individual crop or livestock data, enabling site-specific decisions to optimize resource use, productivity, profitability, and sustainability.

Precision agriculture, often known as precision farming, is a state-of-the-art kind of crop management that uses technology to monitor, evaluate, and maximize resource utilization in agricultural operations, as shown in Figure 12.15. Precision farming enables farmers to increase crop yields, decrease waste, improve sustainability, and make better decisions by utilizing data, sensors, GPS, and other technologies.

FIGURE 12.15 Precision farming [19].

Characteristics

- Data is gathered in real time to track weather patterns, crop conditions, and soil health.
- Makes more accurate judgments, which increases production and decreases waste.
- Often makes use of IoT sensors, drones, and self-governing equipment.

Examples: Using GPS to monitor irrigation requirements and manage pesticide application.

12.2.12 Urban Farming

Definition: Urban farming that frequently employs cutting-edge techniques including community gardens, rooftop gardens, and vertical farming.

The activity of cultivating food in or near cities using small-scale agricultural methods in urban settings is known as urban farming, as displayed in Figure 12.16. In order to grow crops, rear animals, and generate food for local consumption, this type of farming uses existing places like roofs, vacant lots, community gardens, and even inside spaces. From tiny container gardens to bigger, more commercial urban farms, urban farming may take many different shapes.

Characteristics

- Limited space and dependence on other food-growing techniques.
- Puts an emphasis on lowering carbon emissions and producing food locally.
- Can contribute to urban food security.

Examples: Producing fruits, vegetables, and herbs in urban areas.

FIGURE 12.16 Urban farming [20].

12.3 PROPOSED METHOD FOR IMPLEMENTATION OF PRECISION FARMING

In order to achieve optimized yield this method proposes a technique in farming that is capable of integrating the latest technologies in farming. The method is commonly known as hydroponics. In this method plants are cultivated without soil, inducing rich nutrients to water further carried by roots.
The major components of hydroponics are:

1. Monitoring, sensing, and actuation devices.
2. Analysis of captured data through sensors.
3. Processing unit aiding to automation.

Sensing elements such as DHT11, LDR, and MQ-3 for monitoring temperature, humidity, pH, CO_2, and ambient light are embedded to achieve precision in farming. These sensors are headed as Nutrient Sensors. Finally, since the crops are cultivated in water a setup for water level monitoring is also required. The setup includes an automated pump that checks and auto-pumps the water based on requirements. Interconnecting the sensors in a hydroponic system, i.e., agribot using Raspberry pi, is designed and analyzed for its performance.

The data captured through the monitoring process is further analyzed using a machine learning algorithm such as Deep Neural Networks (DNN). This analysis further helps to optimize nutrient delivery, light control, irrigation, etc.

The process flow of this work is depicted in Figure 12.17. The usage of Raspberry Pi3 is effective for this proposed work since automation and control are possible along with monitoring. The sensor is utilizing the Raspberry Pi3 in a hydroponics project. Including The Raspberry Pi3 should be connected to environmental sensors including light, temperature, and humidity sensors (DHT11).

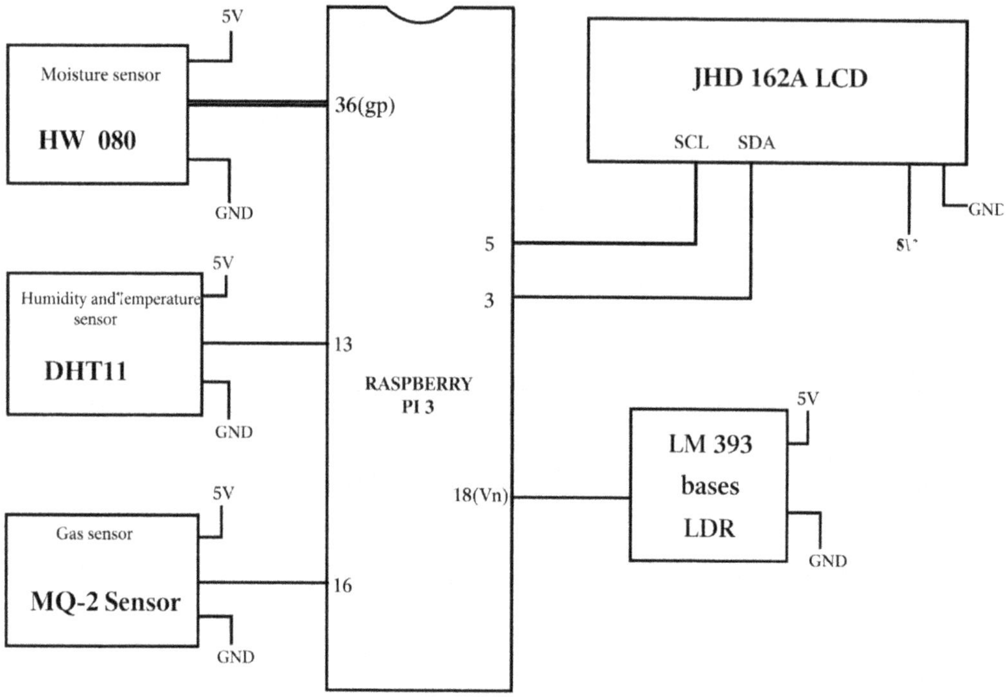

FIGURE 12.17 Implementation of hydroponic system using Raspberry Pi3.

These sensors' data can be read by the Raspberry PI3 and sent for additional processing. Use the Raspberry Pi3 to implement data logging features that will allow you to record sensor data over time. Making educated decisions regarding system optimization and studying trends and patterns can both benefit from this historical data. The RASPBERRY Pi3 is interfaced with a DHT11 sensor that detects the surrounding temperature and humidity. Using the measured temperature and humidity, a DC motor-driven propeller fan and automated water sprayer are activated. An analog moisture sensor measures the amount of water present; when the value is low, a relay-controlled water pump moves water from the external tank into the main tank. In hydroponics projects, gas sensors are a useful tool for monitoring and managing the air quality in the growing environment [21].

Implementing an IoT-based hydroponic system utilizing DNNs involves a multifaceted approach, integrating hardware, software, and data science methodologies. At the core of this implementation lies the deployment of various sensors strategically positioned within the hydroponic environment to capture vital parameters like temperature, humidity, pH levels, nutrient concentration, and light intensity. These sensors serve as the data acquisition nodes, continuously collecting real-time information crucial for monitoring and regulating the hydroponic system's conditions [22].

Figure 12.18 shows that data so captured is wirelessly transmitted and stored in the cloud; the thinkspeak cloud platform is used for this purpose. Later the data is analyzed using DNN. DNN is applied after data preprocessing through cleaning, which cleanses and removes the outliers. An alternative process of training using an appropriate dataset supports decision autonomously. The decisions leverage to adjust nutrient concentration and adaptation to environmental parameters catering to overall growth and health optimization in plants.

The system designed using DNN acts like a backbone for a local hydroponics setup and the data gathered by the cloud through sensors are fed to the automation system to generate final decisions and further actuators are triggered accordingly to maintain overall balance in the environment for optimized growth. The user is also given access to manually control the actuation. The metrics related to growth can be observed in real time through the user interface which has a personal login, as shown in Figures 12.19 and 12.20, and based on user needs control commands can be triggered. In this work, DNN is chosen since it is multifaceted, robust, complex, and interpretable.

While the benefits of integrating DNNs into IoT-based hydroponic systems are abundant, challenges such as data security, computational complexity, and model interpretability must be addressed meticulously. Nonetheless, the potential for revolutionizing modern agriculture through the synergy of IoT and AI is immense, promising sustainable, data-driven solutions to feed a growing global population while conserving precious resources [23].

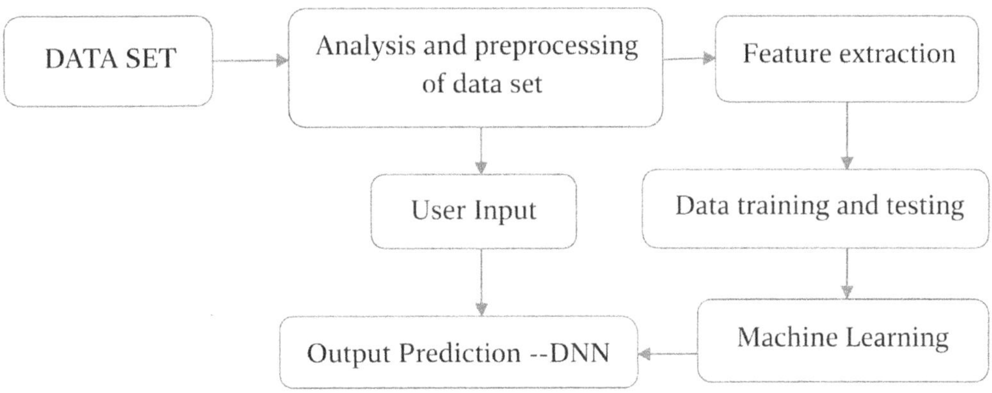

FIGURE 12.18 Implementation of IoT-based hydroponic system utilizing deep neural networks.

FIGURE 12.19 Illustration of User Interface to visualize the status of the crop.

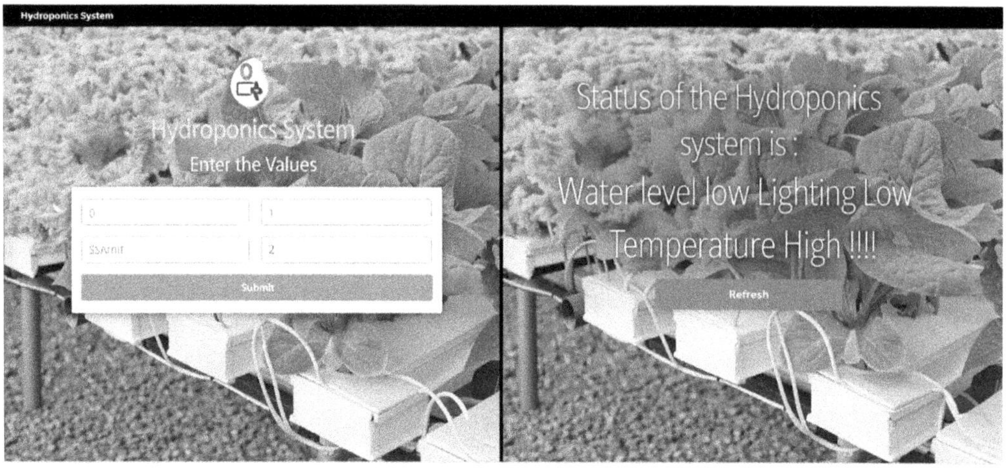

FIGURE 12.20 Visualization of sensor values through UI.

The sensed values are the inputs to the thingspeak cloud and outputs are the conditions to be set up for better yield generation. The cloud used is ThingSpeak. ThingSpeak takes data and updates it every 10 seconds. Then these values are sent and input to DNN; then these values are directly visible on the screen and according to the values, the problem is predicted [24].

Figure 12.21 depicts the nature of the cultivating environment in terms of humidity, temperatures, pH, and ambient light for a particular time and date. The variations in the graph show changes in particular parameters, and only pH consistency is observed, which indicates the nature of water is neutral, and only based on the crop it must be changed [25].

12.4 CONCLUSION

Utilizing technology, data analysis, and automation to maximize agricultural operations, precision agriculture is regarded as one of the most sophisticated and successful types of contemporary farming. It enables farmers to administer pesticides, fertilizer, and water exactly where and when they are required, increasing crop yields while lowering input costs and waste. Precision agriculture is more sustainable and resource-efficient than conventional, organic, hydroponic, or intensive agricultural techniques. Farmers may make better decisions by using real-time crop health, soil, and weather monitoring technologies including GPS mapping, soil sensors, drones, and satellite photography. The long-term advantages in terms of cost savings, environmental preservation, and productivity make it a better option, even though the early setup costs may be substantial. Additionally, it aids farmers in adjusting to environmental difficulties and climate change, making it a progressive approach to ensuring food security worldwide. Therefore, when efficiency, sustainability, and profitability are taken into consideration, precision agriculture is the most effective agricultural technique.

12.5 FUTURE RESEARCH DIRECTIONS

The goal of future hydroponic farming research should be to make the system more accessible, economical, and effective for producers of all sizes. To help smallholder and urban farmers, one important avenue is the creation of affordable hydroponic systems with locally accessible or repurposed components. Along with developing automated systems that can modify nutrients in real time, it is also essential to optimize nutrient solutions for various crops and growth phases. Including renewable energy sources, such as wind or solar, may improve sustainability and save operational costs. Research is also required to create crop kinds that are especially adapted to hydroponic environments, guaranteeing greater yields. Resource management, pest control, and system monitoring may all be further enhanced by the application of smart technologies like IoT, AI, and remote sensing. To improve influence on the environment, greater attention must be paid to biological pest management and water recycling techniques. Furthermore, combining hydroponics with urban and vertical farming offers potential ways to produce food in places with limited space. Lastly, in order to comprehend acceptance constraints and advance hydroponics as a practical agricultural technique globally, socioeconomic research and supporting legislation are crucial.

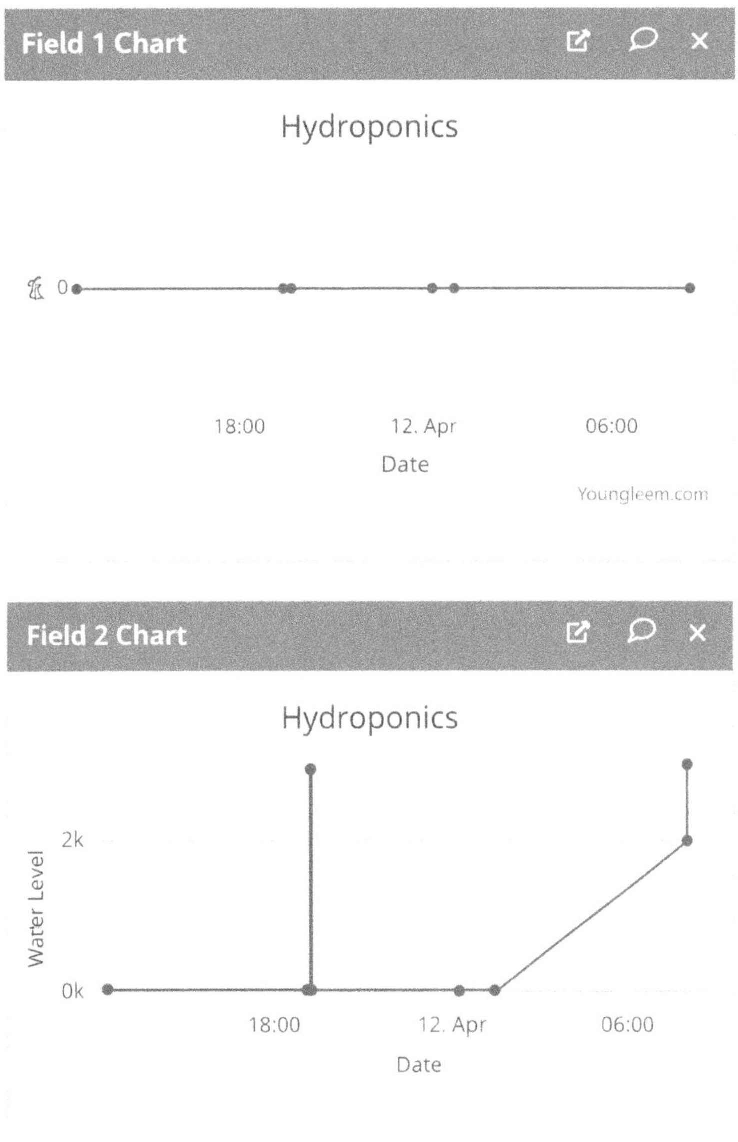

FIGURE 12.21 Results obtained through Thinkspeak after analysis of the data for humidity, temperature, pH, light intensity, and water level.

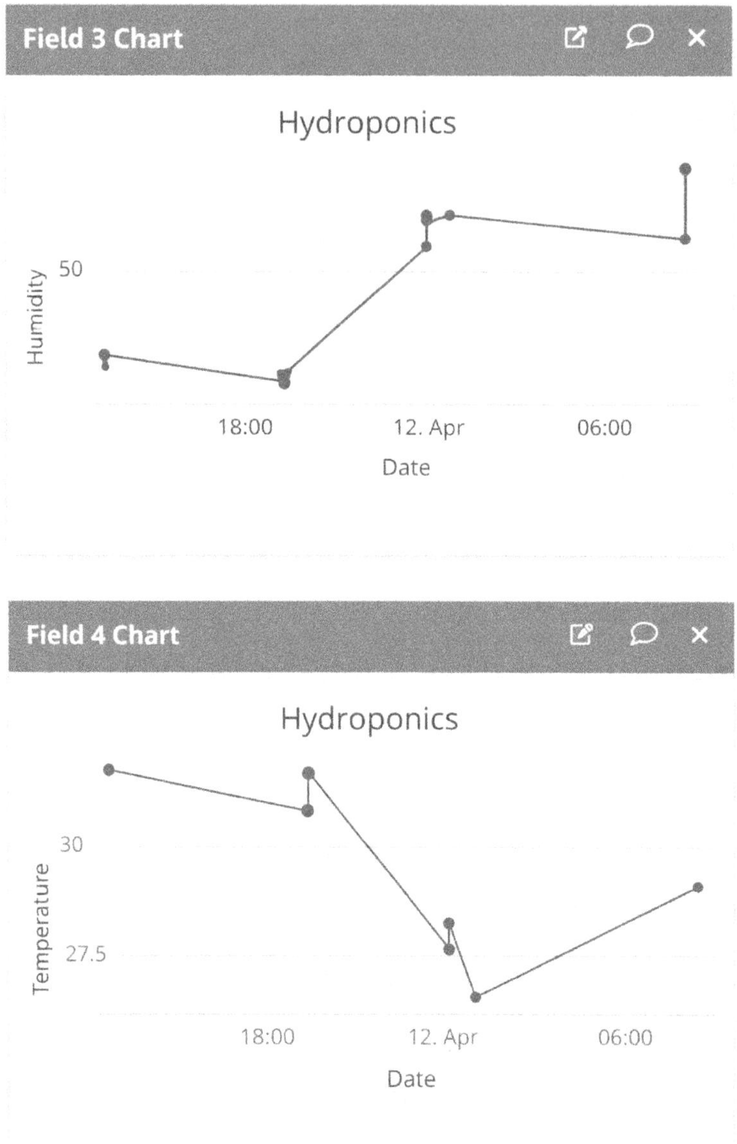

FIGURE 12.21 Continued

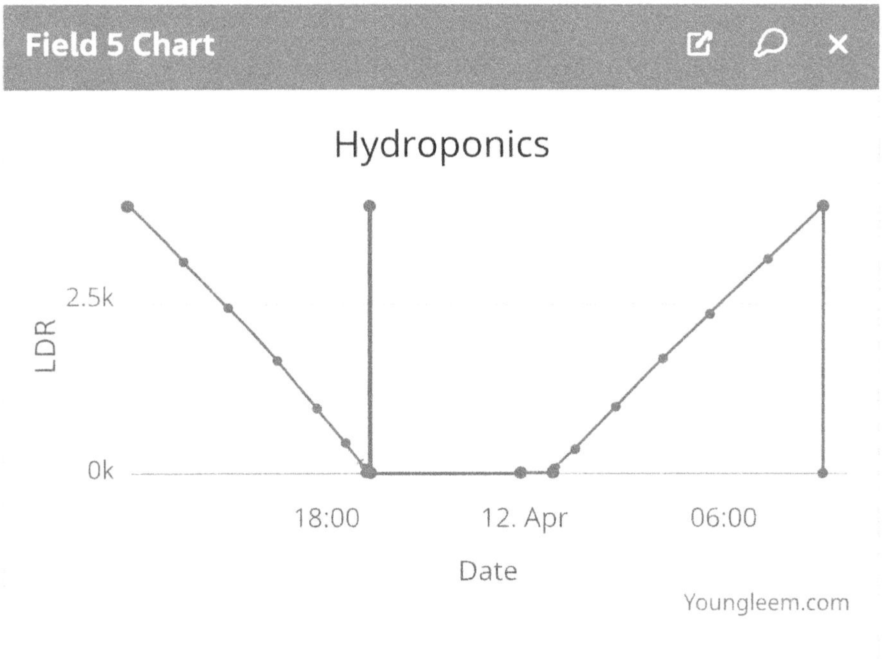

FIGURE 12.21 Continued

REFERENCES

[1] S. Chand and S. Kumar, "Structural Transformation and the Declining Role of Agriculture in India's GDP," *Economic and Political Weekly*, vol. 55, no. 12, pp. 45–53, Mar. 2020.

[2] Sudha Ellison Mathe, Mamatha Bandaru, Hari Kishan Kondaveeti, Suseela Vappangi, and G. Sanjiv Rao, "A survey of agriculture applications utilizing Raspberry Pi," *International Conference on Innovative Trends in Information Technology (ICITIIT)*. 2022. 978-1-6654-0670-3/22/$31.00 ©2022 IEEE | Doi: 10.1109/ICITIIT54346.2022.9744152

[3] R. Chand, S. Parappurathu, and S. Kumar, "Agricultural Development and Structural Transformation in India," *Economic and Political Weekly*, vol. 50, no. 24, pp. 55–62, Jun. 2015.

[4] R. Barman, R. Singh, and R. N. Padaria, "Agriculture 4.0: The Indian narrative – Tracing its roots and growth," *Food and Human*, vol. 5, no. 100608, pp. 1–12, 2025, doi: 10.1016/j.foohum.2025.100608.

[5] Z. Muhammad, M. A. A. M. Hafez, N. A. M. Leh, Z. M. Yusoff, and S. A. Hamid, "Smart agriculture using Internet of Things with Raspberry Pi," *Food and Human*, vol. 17, no. 1, pp. 45–53, 2025.

[6] Ahmed J.O Ahmed, Ahmed Babiker, Ali Elhag, and Mohamed Drar, "Real-time agricultural monitoring with Agrobot: A Raspberry Pi and YOLO based solution," *International Conference on Computer and Applications (ICCA)*. 2023. Doi 10.1109/ICCA59364.2023.10401368

[7] P. Birthal, A. Negi, and S. Kumar, "Farm Mechanization and Agricultural Productivity in India: Policy and Evidence," *Agricultural Economics Research Review*, vol. 29, no. 2, pp. 167–178, Dec. 2016.

[8] P. L. Pingali, "Green Revolution: Impacts, limits, and the path ahead," *Proceedings of the National Academy of Sciences of the United States of America*, vol. 109, no. 31, pp. 12302–12308, 2012.

[9] R. Singh and P. Kumar, "Farm Mechanization and Its Impact on Agricultural Productivity in India," *Journal of Agricultural Engineering*, vol. 52, no. 3, pp. 15–24, Sep. 2015.

[10] A. Kumar, R. Sharma, and P. Gupta, "Process optimization in viticulture using functionalized biomimetic Hydroxyapatite (HAB) nanostructured particles, spectrographic analysis, and drone spraying systems," *Journal of Nanoscience and Nanotechnology*, vol. 26, no. 2, pp. 145–156, Feb. 2024.

[11] D. Tilman, K. G. Cassman, P. A. Matson, R. Naylor, and S. Polasky, "Agricultural sustainability and intensive production practices," *Nature*, vol. 418, no. 6898, pp. 671–677, 2002.
[12] B. Nare and S. Singh, "Mechanization in Potato Cultivation," *Indian Horticulture*, vol. 64, no. 6, pp. 315–397, 2019.
[13] S. Bhardwaj and R. Sharma, "Enhancing Potato Farming through Mechanization in India: A Review," *Agricultural Sciences: Techniques and Innovations*, vol. 1, pp. 53–66, Apr. 2025.
[14] P. D. Shinde and A. S. Mali, "Hydroponic Solution for Soilless Agriculture with Advanced Techniques and Smart Perspectives," in *Proc. Int. Conf. on Electronics Engineering and Technology*, Kolhapur, India, 2024.
[15] H. M. Resh, *Hydroponic Food Production: A Definitive Guidebook for the Advanced Home Gardener and the Commercial Hydroponic Grower*, Boca Raton, FL, USA: CRC Press, 2012.
[16] S. R. Patil, A. K. Sharma, and P. D. Shinde, "IoT Integrated Hydroponic System for Precision Agriculture," in *Proc. IEEE International Conference on Electronics and Sustainable Agriculture Systems (ICESAS)*, Pune, India, pp. 112–118, Mar. 2024.
[17] M. S. Reddy and S. K. Rao, "IoTEnabled Hydroponic Farming System for Urban Agriculture,"- *IEEE Sensors Journal*, vol. 23, no. 5, pp. 1123–1132, Mar.2023.
[18] H. Zhang, Y. Li, and J. Chen, "AIDriven Nutrient Optimization in Hydroponic and Permaculture Systems,"- *IEEE Access*, vol. 11, pp. 45678–45690, Jun. 2023.
[19] C. R. Mehta, K. K. Singh, B. S. Gholap, and R. Patle, "Status of Mechanization of Potato Cultivation in India," *Agricultural Engineering Today*, vol. 42, no. 2, pp. 37–44, 2018.
[20] M. ElSayed-, and H. Ali, "Two rooftop farming concepts for residential buildings: Shifting towards sustainable communities—An exercise on the case of Asmarat with necessary readaptation of its architectural rooftop design,"- *International Journal of Sustainable Architecture and Urban Design*, vol. 12, no. 3, pp. 145–162, Mar.2024.
[21] S. Rajendran, T. Domalachenpa, H. Arora, P. Li, A. Sharma, and G. Rajauria, "Hydroponics: Exploring innovative sustainable technologies and applications across crop production, with emphasis on potato minituber cultivation,"- *Heliyon*, vol. 10, no. 5, pp. e123456, 2024.
[22] I. A. Lakhiar, J. Gao, T. N. Syed, F. A. Chandio, and N. A. Buttar, "Modern plant cultivation technologies in agriculture under controlled environment: A review on aeroponics," *Agricultural Engineering International: CIGR Journal*, vol. 21, no. 1, pp. 1–15, 2019.
[23] G. Rajaseger, K. L. Chan, K. Yee Tan, S. Ramasamy, M. C. Khin, A. Amaladoss, and P. Kadamb Haribhai, "Hydroponics: Current trends in sustainable crop production," *Bioinformation*, vol. 19, no. 9, pp. 925–938. 2023 (Sep 30). Doi: 10.6026/97320630019925. PMID: 37928497; PMCID: PMC10625363.
[24] K. K. Y. Shin, T. P. Ping, M. G. B. Ling, C. Chee Jiun, and N. A. B. Bolhassan, "SMART GROW – Low-cost automated hydroponic system for urban farming," *HardwareX*, vol. 17, pp. e00498, 2023 (Dec 9). Doi: 10.1016/j.ohx.2023.e00498. PMID: 38130490; PMCID: PMC10731670.
[25] A. Soussi, E. Zero, R. Sacile, D. Trinchero, and M. Fossa, "Smart sensors and smart data for precision agriculture: A review," *Sensors* (Basel), vol. 24, no. 8, pp. 2647, 2024 (Apr 21). Doi: 10.3390/s24082647. PMID: 38676264; PMCID: PMC11053448.

13 Classification of Pests and Diseases in Maize Crop Using Hybrid Model (YOLOV11+VGG16)

Sharmiladevi S, Vigrahala Rajeshri, and G L K Vani Pradhayi

13.1 INTRODUCTION

Maize is one of the staple crops that make a huge contribution to the food security and economic stability of many developing and developed countries. However, its production is still massively at stake because of wide range of pests and diseases that may cause a great yield loss, decrease crop quality, and increase the cost of cultivation [1, 2, 3, 4, 5, 6]. The traditional detection methods that are mostly dependent on manual inspection are quite inefficient since they are time-consuming, require a lot of labor, and are quite vulnerable to mistakes made by people, especially in rural areas or areas that are resource-constrained where the number of experts for agricultural support is quite limited [1, 2, 7, 8, 9]. The lack of appropriate and accurate detection methods on demand allows the infections to spread fast, and thus results in the occurrence of widespread crop failures and thus an economic crisis for farmers.

To tackle those difficulties, achievements in artificial intelligence (AI) – deep learning in particular – have proved to be able to greatly facilitate the work of automating the process of pest and disease identification. Early machine learning (ML) models like Support Vector Machines (SVM), Random Forests (RF), and k-Nearest Neighbors (KNN) tried to provide some level of automation; however, they were not able to handle the scale and complex visual patterns found in the real-world agricultural datasets [10, 11]. Today, deep learning architectures such as Convolutional Neural Networks (CNNs) have already overtaken traditional methods by directly learning low- to high-level image features. Besides that, object detection frameworks like You Only Look Once (YOLO) have also become available to perform multi-class detection from static images in just one pass, and, therefore, they are suitable for agricultural applications which request both accuracy and speed [5, 12, 13, 14, 15, 16]

To address the limitations of using single models only, this study proposes a combination deep learning technique that integrates YOLOV11 for object detection and VGG16 for image classification. The responsibility of YOLOv11 is to find the pests and localize the areas of the crop that are infected by disease in pictures of maize. VGG16 is also employed to decide upon the categories of the detected objects. This hybrid model utilizes the object localization capabilities of YOLO and the classification accuracy of VGG16 to achieve the best possible performance. A comparative evaluation against baseline models such as CNN, VGG16, and YOLOV11 reveals that the hybrid system not only significantly boosts detection accuracy but also helps to reduce misclassification.

The purpose of this study is to analyze the way AI works and mainly how deep learning can be used for finding maize crop pests and diseases. Section 13.2 (Literature Review) covers some of the existing methods to the problem focusing on traditional ML models, deep learning architectures, and object detection frameworks. Section 13.3 (Proposed Methodology) explains the idea of the

DOI: 10.1201/9781003613510-13

hybrid model, which is a network composed of YOLOv11 and VGG16 to make the detection accuracy better. Section 13.4 (Experimental Setup and Dataset) explains in detail the hardware and software configurations for model training and evaluation. It concentrates on the topic of data collection, preprocessing techniques, and class distribution to ensure robust model training. Section 13.5 (Experimental Results) consists of tables and diagrams showing the respective performances of different models and presenting a solid argument to support the hybrid model approach and explanation of each result; it also highlights the most important results of the experiment by presenting them in an easy way. On the other hand, Section 13.6 (Discussion) takes a deeper look at the obtained results by underlining the strong and weak aspects of the results and gives possible future improvements. Section 13.7 (Conclusion) makes note of the study's successfulness by saying that the pest and disease detection for maize crops was greatly improved using the hybrid model. It also investigates the optimizations that can be made in terms of computational performance and data expansion to achieve high accuracy.

13.2 RELATED WORK

The identification of crop pests and diseases has been transformed from the use of traditional manual inspection methods to the utilization of Al-driven advanced solutions. Initially, these methods involved manual observation that required much time and was often accompanied by mistakes. Then the era of ML came and introduced automated classification which was done by algorithms like SVM, RF, KNN, and Artificial Neural Network (ANN), which had a positive impact on resolving the detection and accuracy issues. Yet, ML models faced complications with big agricultural datasets as well as multifaceted visual features, which in turn led to the adoption of methods such as computer vision architectures like CNNs and transformer models, etc. Object detection models like YOLO and Faster Regional-based-Convolutional Neural Networks (R-CNN) also increased the real-time detection of pests. Hybrid Al-driven innovations such as CNN-LSTM hybrid models, Few-Shot Learning (FSL), and Attention-GAN are the most recent developments that integrate the best aspects of different areas.

Pest and disease detection in crops has progressed from manual inspections to AI-based methods. Traditional ML models such as SVM, RF, KNN, and ANN improved accuracy yet struggled with scalability, complex feature extraction, and computational efficiency [1, 10, 11, 7, 17, 18, 19]. For example, SVM performed well on small datasets but poorly on high-dimensional data [10], RF achieved moderate accuracy but was resource-heavy [7], and KNN was computationally expensive for large-scale use [10]. These limitations led to the adoption of deep learning (DL).

CNNs like AlexNet, ResNet, GoogleNet, DenseNet, VGG16, and EfficientNetV2 attained high performance in disease classification [12, 20, 21, 22]. Transfer learning approaches (e.g., VGG16 and EfficientNetV2) boosted accuracy for grape, tomato, and other crops [22]. ResNet-based variants improved computational efficiency [21], while self-supervised learning (SSL) helped address dataset scarcity [23].

Object detection models such as YOLO and Faster R-CNN enhanced real-time pest detection, surpassing classical ML and CNNs in speed and accuracy [5, 16, 24, 25]. Variants like Maize-YOLO [5], CTR_YOLOv5n [26], and YOLOv8 [27] achieved high detection precision, though challenges remain with generalization and environmental variability.

Hybrid models, combining CNNs with LSTMs, Few-Shot Learning (FSL), Attention-GAN, and transformer-based architectures, further improved disease forecasting and detection under limited data conditions [28, 29, 30, 31].

Finally, AI-driven smart agriculture integrates IoT sensors, UAVs, CNNs, YOLO, and Faster R-CNN for large-scale, real-time monitoring, demonstrating detection rates up to 99.63% [2, 32, 33, 34]. These systems optimize pesticide use and strengthen crop management by enabling early, accurate detection of pests and diseases.

13.3 PROPOSED METHODOLOGY

This section describes about the maize pest and disease identification through the proposed hybrid model. It presents the hybrid model (YOLOv11 + VGG16) and explains its usage for both classification and object detection. VGG16 is a deep learning model specializing in feature extraction and classification, which results in high accuracy in image recognition applications. Object detection is one of the features of YOLOv11 and it is combined with VGG16 in the Hybrid Model (as shown in Figure 13.1) to increase accuracy in both detection and classification. The accuracy of these models

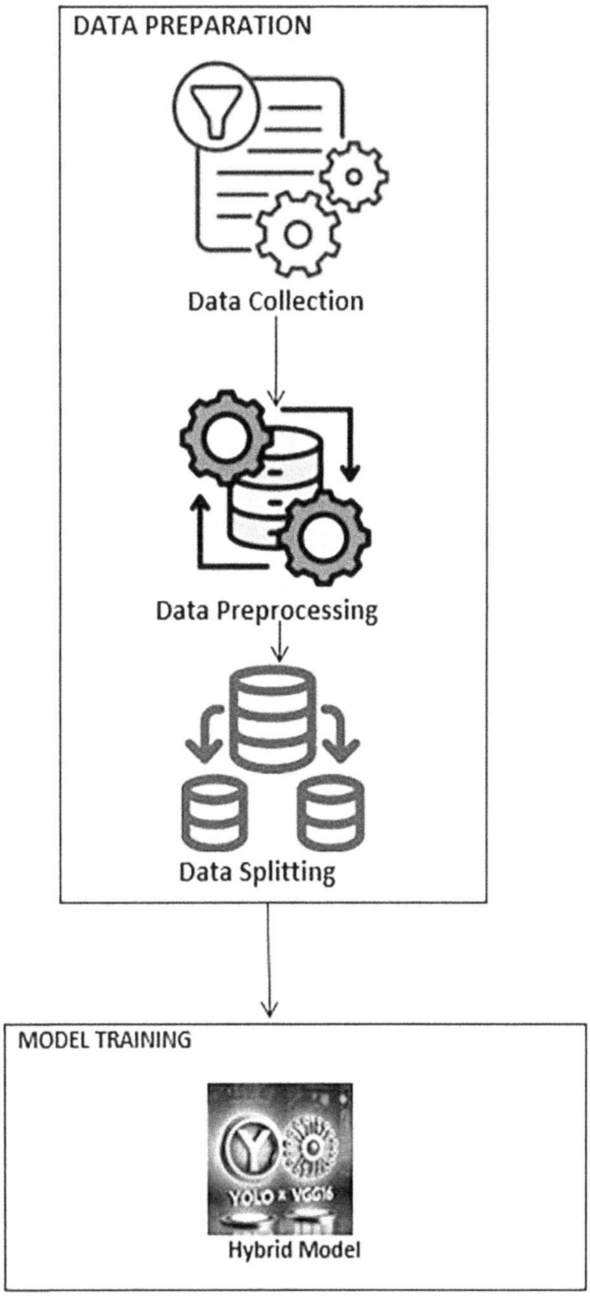

FIGURE 13.1 Proposed hybrid model (YOLOv11 + VGG16).

is calculated by employing different performance metrics like accuracy and confusion matrices, which are used to find the most reliable approach for maize pest and disease identification.

13.3.1 Data Preparation and Data Preprocessing

In this stage, data is gathered and processed. The aim is to have the model get high-quality inputs for training. The whole process of this application includes three stages. Data acquisition is the first step, wherein data such as maize crop images are collected. Some of these pictures represents different conditions that are likely to cause damage, such as pest infestations (e.g., fall armyworm, grasshopper, and leaf beetle), fungal infections (e.g., leaf spot), and viral diseases (e.g., maize streak virus).

In image preprocessing step, image resizing, normalization, and augmentation are carried out, which are crucial for developing hybrid model functionality. The final step is dividing the dataset into three parts that represent the training, validation, and testing to properly train the model and accurately evaluate its performance.

13.3.2 Hybrid Model Training

In the Hybrid Model, YOLOv11 will do the object detection to determine the regions for pests or diseases in the image. Then VGG16 will classify the objects which contributes to precise identification of the pests and diseases. The combination of both the models is very useful because it is capable of not only classifying objects correctly but also seeing the surrounding objects.

The maize pest and disease detection hybrid model integrates YOLO with a CNN transfer learning architecture VGG16 where YOLO is employed for object detection and VGG16 for classification. The YOLOv11 model has been used to identify and locate pests and diseases in maize images provided as input and returns outputs by enclosing them in rectangles. Subsequently, the isolated components are transferred to a VGG16 model that is used for classifying pest or disease in maize crop so that it can be identified correctly. The model goes through a test dataset, containing images of pests and diseases of maize crop and predicts corresponding classes based on confidence scores obtained. It forecasts the impacted regions, provides confidence scores, and graphically outputs result with labeled predictions and detection confidence.

13.4 EXPERIMENTAL SETUP AND DATASET

The efficiency of the deep-learning models, particularly those based on CNNs and YOLO, which work with object detection frameworks as the core, is highly dependent on the hardware and software functions of the tools. The subsequent subsection will describe the necessary hardware and software configurations that should be provided for optimal model training, inference, and dataset processing.

The maize pest and disease detection software and hardware resources are both critical for efficient functioning. The NVIDIA T4 GPUs are instrumental in speeding up the process of the deep learning model in terms of training and inference. Moreover, for the purpose of huge number of data with complex augmentation techniques, a 40 GB RAM is used. On the other side, the software element comprises the use of a Windows 10 operating system that is compatible with deep learning framework and development tools.

13.4.1 Dataset

The dataset used in the classification of maize crop pests and diseases was made with a focus on the numerous issues associated with maize. It contains six distinct classes, each representing either a distinct pest, disease, or healthy maize crops. This dataset is used to train the CNN-based classification models and the YOLO-based object detection models, which are meant to detect and properly classify the pests and diseases that affect maize crops.

The six categories of the dataset are:

1. **Maize Fall Armyworm** – A major pest responsible for the huge damage to the maize crop by grazing the leaf and resulting in the loss of yield.
2. **Maize Grasshopper** – Grasshopper is very common globally. This insect can even lead to a decrease in plant strength by eating maize leaves.
3. **Maize Healthy** – The word "healthy" refers to the images of plants, which are those left standing without any indications of pest or disease infestation.
4. **Maize Leaf Beetle** – It is the case of the insect pest that creates holes in the maize leaf and leads to weakening of the leaf.
5. **Maize Leaf Spot** – A fungal disease that causes appearance of brown spots on the leaves and, in turn, results in decreased photosynthesis.
6. **Maize Streak Virus** – It is a viral disease that is shown by the streak-like yellowish-green marks on the leaves, which is seen as the reason for the extensive damage and the decline in productivity of the plants.

13.4.1.1 Data Collection and Annotation

To have a more reliable dataset, first, images were collected from the Kaggle which is of the size 4,389. Since few classes contain less number of images, Roboflow is used for augmentation and to increase the dataset size and also to balance all classes in the dataset. In this Roboflow website, YOLO model is used to annotate the images.

The platform provided an easy-to-use interface through which the researcher could manually draw as many bounding boxes as the number of diseases or pests within images to tag them correctly. After exporting the annotations to YOLO format, a text file was produced with a single line in relation to each object recognized in the image, where the object is composed of a class ID and the normalized coordinates of the bounding box.

Finally, the annotated datasets were subsequently used in the training of the YOLO model for object classification and detection.

13.4.1.2 Augmentation and Data Preprocessing

Roboflow's native augmentation features were employed in implementing data augmentation methods for enhancing model robustness and generalization. Indeed, augmentation plays a vital role in object detection tasks, as it allows the model to recognize objects regardless of the orientation, brightness, and background settings. The following are the data augmentation strategies applied:

1. **Rotation** – Randomly rotating images with small angles to enable the model to identify pests and diseases from different directions.
2. **Scaling and Resizing** – Similarly to easily process the entire dataset, the image sizes are normalized to achieve the same input size parameters.
3. **Brightness and Contrast Adjustments** – By simulating the different lighting, one can enhance the model's performance under different climatic situations.
4. **Flipping** – Horizontal and vertical flipping is used to generate variations in the dataset.
5. **Cropping and Padding** – Random cropping of image areas and padding of even the aspect ratio to be preserved.

Because of these augmentations, the model gets dataset variability, which makes the model avoid overfitting to specific patterns that are seen only during training.

13.4.1.3 Dataset Split Strategy

In order to run an equal training and evaluation process, the dataset was segmented into three sets as follows:

Training Set (70%) – The majority of the images were divided into training so that the model would be able to teach the network patterns more effectively.

Validation Set (10%) – This subset was used to fine-tune the model's hyperparameters and avoid overfitting. It is a set of data that helps to optimize the model.

Test Set (20%) – Another portion of the pictures that were left hidden during the training process and never used for model performance testing.

Through the Roboflow annotation tools and augmentation methods, the dataset was enhanced to improve the training of CNN-based classification models like VGG16 and YOLO-based object detection models. The precise annotations, various augmentation techniques, and clean dataset splits give the models a boost in terms of accuracy and efficiency, which in turn makes them more trustworthy for efficient maize pest and disease detection applications.

13.5 EXPERIMENTAL RESULT ANALYSIS

Maize pest and disease detection experimental results using CNN, VGG16, YOLOv11, and Hybrid Model (YOLOv11 + VGG16) are presented in this section. The models are assessed using various performance metrices such as the accuracy, confusion matrices, precision-recall curves, and F1-confidence curves. The different models used in this study are explained in the following sections along with their performances.

13.5.1 CNN

CNN is designed for image classification and feature extraction [12]. It is widely used in computer vision tasks because of its ability to learn spatial hierarchies from images through convolutional layers. CNNs have the ability to find edges, textures, and shapes in a very short time by themselves [35]. It is also good for recognizing the object in the image. In contrast to the traditional ML models such as SVM, Decision Trees, k-NN, CNNs do not require manual feature extraction because they find it from the raw image data.

In this study, the CNN model is used for comparing results obtained by hybrid model with respect to classifying the pests and diseases of the maize crop. A dataset of pictures of the maize crop is trained for telling apart the different pests and diseases. This model consists of feature identification convolutional layers, feature size reduction pooling layers, and fully connected model layers for choices made at the end. The SoftMax activation function is applied in the output layer to allocate probabilities to the classes.

13.5.1.1 Performance of CNN Model

The goal of this section is to evaluate the accuracy and loss trends of the CNN model when used for the detection of maize pests and diseases, to eventually classify them. The model is examined for its ability to take the features of the dataset and learn how to adapt them over time.

Figure 13.2 illustrates the CNN model exhibiting a continuous growth and it can be inferred that the CNN model has been learning properly. The close resemblance in the training as well as in the classification validation accuracy guarantees that the model can efficiently recognize the maize pests and diseases.

Figure 13.3 presents the trends in training and validation losses. At first, the validation loss is extremely high but as time moves on it goes down. This means that the model is indeed learning well and decreasing mistakes step by step, therefore ensuring the exact classification of maize pests and diseases.

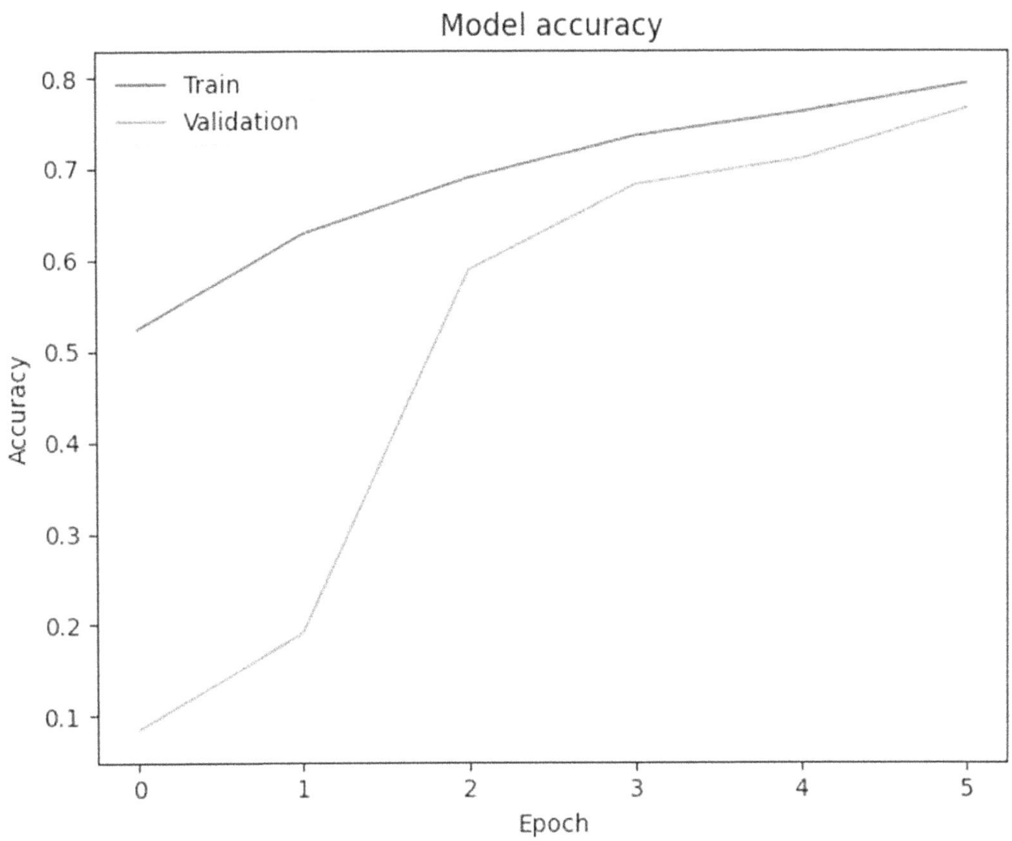

FIGURE 13.2 Accuracy graph of CNN.

13.5.2 VGG16

VGG16 is a deep convolutional network (CNN) that assists in image classification. It has 16 layers, which has 3 × 3 filters and 3 fully associated layers with 13 convolutional layers, making this visually a strong model for recognition tasks [36]. The Rectified Linear Unit (ReLU) activation is used in the network to get rid of linearity and max pooling layers are utilized to decrease spatial dimensions, yet retaining the most significant features. Although VGG16 is a heavy-weighted model, it is still one of the most wanted networks during attentive task learning phases because of its superlative feature extraction skills. For this crop pest and disease detection study, the VGG16 model classifies maize crop images into six categories, separating healthy plants from numerous pest and disease conditions. In this way it helps in early diagnosis, and in addition supports farmers in performing timely interventions for the welfare of crops while increasing their yield.

Transfer learning is used to train a maize pest and disease classifier using VGG16, which is a deep CNN model. The resizing, normalization, and labelling of training packet images are all performed to complete this step. The next stage of classification is taking a pre-trained VGG16 model the ImageNet and then by removing the top layers and fine-tuning the last ten layers, respectively. Additionally, a GlobalAverage Pooling2D layer is applied, a Dense-layer with 1,024 neurons, ReLU as an activation function is added for feature learning, and then the last layer Dense (6, softmax) is used for the six categories classification. The model is initially evaluated with various learning rates and the one with the learning rate of $1e^{-5}$ is the best performer. Then it is evaluated on the test

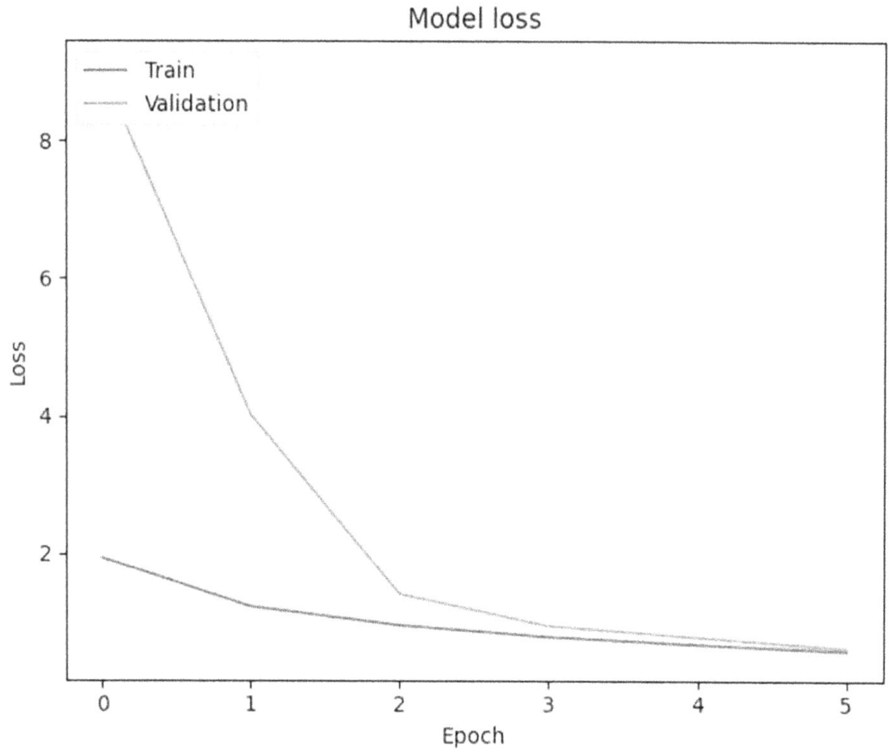

FIGURE 13.3 Loss graph of CNN.

dataset, and, finally, an accuracy score is obtained. In order to make predictions, before testing, the images are sized and normalized to 224x224, then they are put through the model to estimate the probability of each class. The workflow developed in the process of transfer learning is the most notable achievement in VGG16 and it is used in the proposed hybrid model to get better training results.

13.5.2.1 Performance of VGG16 Model

This section presents the accuracy and loss trends of VGG16 model and assesses its strength in classifying maize pests and diseases. The behaviour of this model is evaluated based on its capability to learn and generalize maize crop's diseases. The accuracy curve demonstrates the model's ability to consistently improve throughout its training process. The loss curve shows a reduction in errors over time. These facts are pivotal in assessing the credibility of VGG16 in providing precise classification results.

In Figure 13.4, the VGG16 model's training and validation accuracy curves are provided. The continual increase in accuracy shows the effective and constantly changing nature of the model as it goes through the training process. The performance characteristics of both training and validation sets are very similar which indicates that the model is learning well to recognize maize pest and disease and not memorizing the training set. Hence, it is reliable and well-generalized as a classifier.

The graphical evidence provided in Figure 13.5 adds up to the model's proficiency by demonstrating a consistent decrease in the number of training errors. The closeness between the training loss and validation curves signifies that the model is well-made and it is not overfitting. This sequence emphasizes the VGG16 model's success in the learning process and guarantees an accurate forecast of maize pests and diseases.

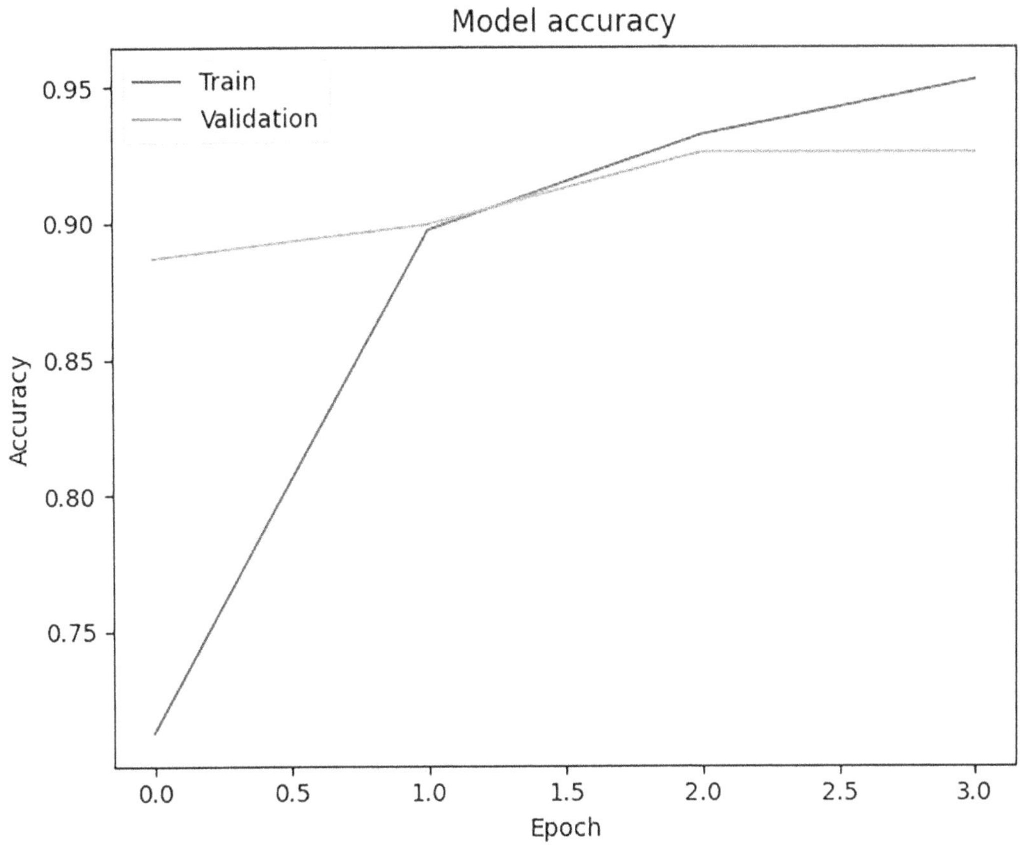

FIGURE 13.4 Accuracy graph of VGG16.

13.5.3 YOLOv11

YOLOv11 is a model for object detection in real time that operates by processing an image in a single forward pass of a neural network. It divides the image into a grid and predicts classes simultaneously, using bounding boxes, class probabilities, and confidence scores. YOLO, which unlike traditional detection models that use region proposals and multiple passes, is fast and efficient, thus making it perfect for real-time applications. Along with speed and accuracy, it employs anchor boxes and non-max suppression techniques to refine the detections YOLOv11 for the detection of maize pests and diseases [37].

YOLOv11 is trained by using the images which were resized to 640x640. Afterward, the best post-training model is assessed based on performance metrics like mean average precision (mAP@50), mAP@50-95, precision, recall, and confidence (F1-score). The trained model is made to work for inference tasks, where the pests and diseases of the maize crop are detected and losses quantified as a score with the help of bounding boxes and confidence scores [38].

13.5.3.1 YOLOv11 Model Performance

The model YOLOv11 excelled in the evaluation metrics, having a mAP@50 value of 0.9303 and a mAP@50-95 of 0.7987, which showed that the accuracy of detection was very high under different Intersection of Union (IOU) thresholds. Besides, this model exhibited a 0.9106 precision rate, a 0.8884 recall rate, and a 0.8994 F1-score, thus ensuring a fair performance since it could detect true

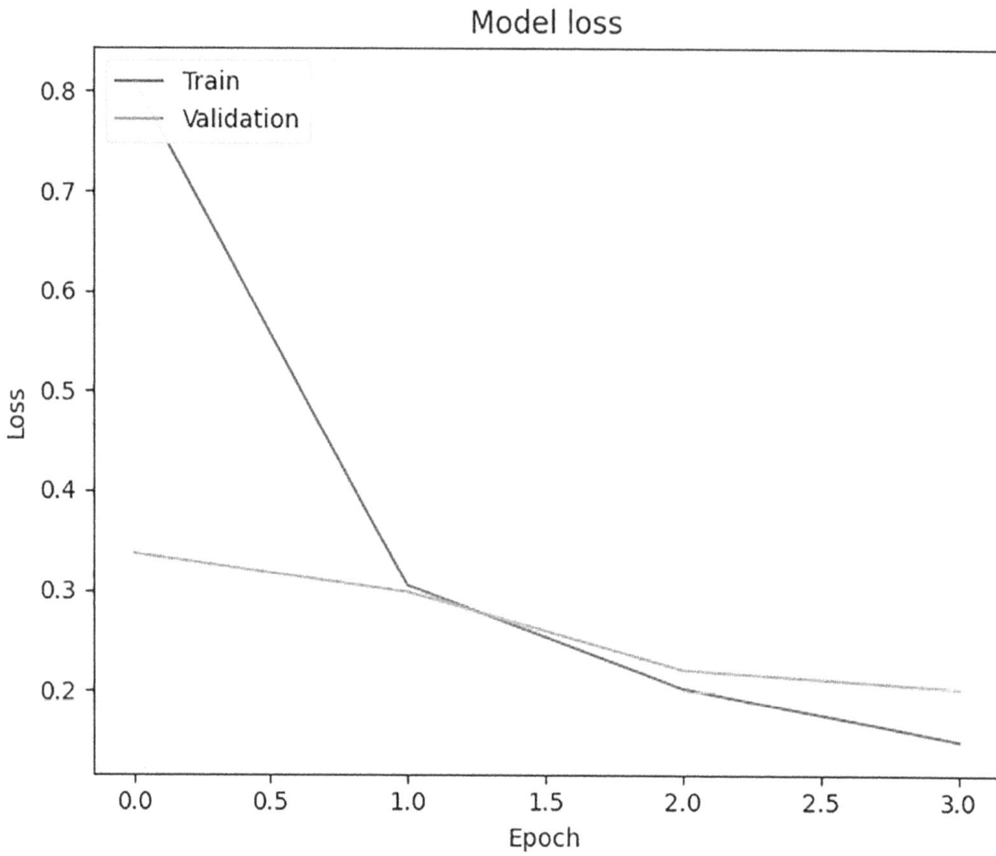

FIGURE 13.5 Loss graph of VGG16.

positives and minimize false detections. An estimated total accuracy of 0.9303 emphasizes once again the efficiency of the model in detecting maize crop's pests and diseases.

Figure 13.6 unveils the normalized confusion matrix of the YOLOv11 model referring to the performance in classifying various categories of maize pests and diseases. The big diagonal values indicate the high classification accuracy of nearly all the classes. Mostly, the highlights included Maize Healthy (0.97), Maize Leaf Beetle (0.95), Maize Leaf Spot (0.91), and Maize Fall Armyworm (0.89) with the least amount of error rates. Whereas Maize Grasshopper (0.86) and Maize Streak Virus (0.89) has less accuracy value. Thus, the model can differentiate between different conditions of crops demonstrating its efficiency.

On the other hand, there are some misclassifications that are seen, and among them, the background class plays an important role. This is because the model at times fails to differentiate between real objects and background, resulting in misclassifications. However, the overall results are in favor of support for YOLOv11's capability of identifying maize pests and diseases, making it one of the options for real-time agricultural disease detection.

Figure 13.7 is a chart that demonstrates a connection between F1-scores and confidence rates of different maize crop conditions and pest categories. Each of the lines represents a class, where the x-axis is confidence and the y-axis is the F1-score.

In many classes, the best F1-scores occur at a 0.5 confidence level and below. That means less accurate classes are at the low level. Among all classes, the best F1-score occurred at 0.487 with

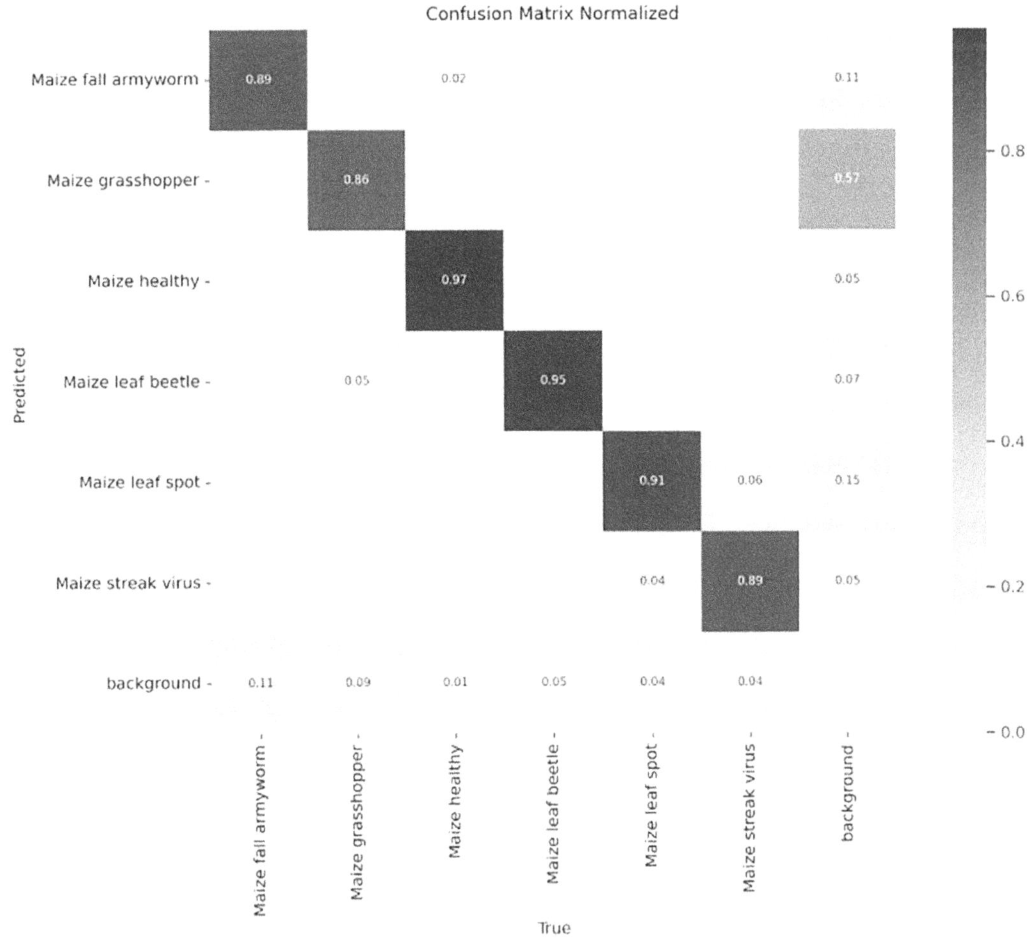

FIGURE 13.6 Confusion matrix of YOLOV11.

a confidence of 0.90. In the case of "Maize streak virus," its shows a very steep decline but only pointing out that only a certain confidence level is a good balance between precision and recall.

The diagram in Figure 13.8 depicts the Precision-Recall curve of YOLOv11 model with the highest mAP@0.5 of 0.930, backing the fact that the detection is strong. Each individual class, e.g., Maize Healthy (0.990), Maize Leaf Beetle (0.973), and Maize Streak Virus (0.952), displays an impressive precision and recall. This model makes a clear distinction between maize pests and diseases, thereby proving its ability to classify accurately.

13.5.4 Hybrid Model (YOLOv11 + VGG16) Performance

The normalized confusion matrix of the hybrid model is provided in Figure 13.9. It can be observed that the model has a high classification accuracy across all maize pest and disease categories. The model provides 100% accuracy for Maize Healthy and Maize Streak Virus, but for other classes such as Maize Leaf Spot (99.2%) and Maize Leaf Beetle (98.78%) it also makes valid predictions. The very few misclassifications demonstrate the model's reliable performance in categorizing the conditions of the maize crop.

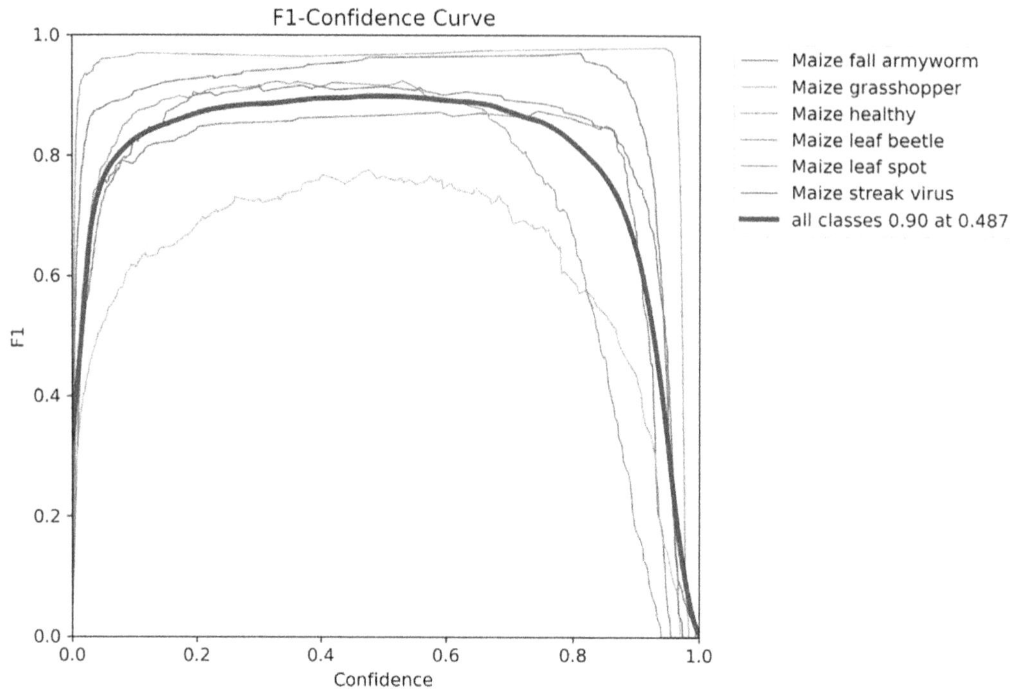

FIGURE 13.7 YOLOV11 F1-confidence curve.

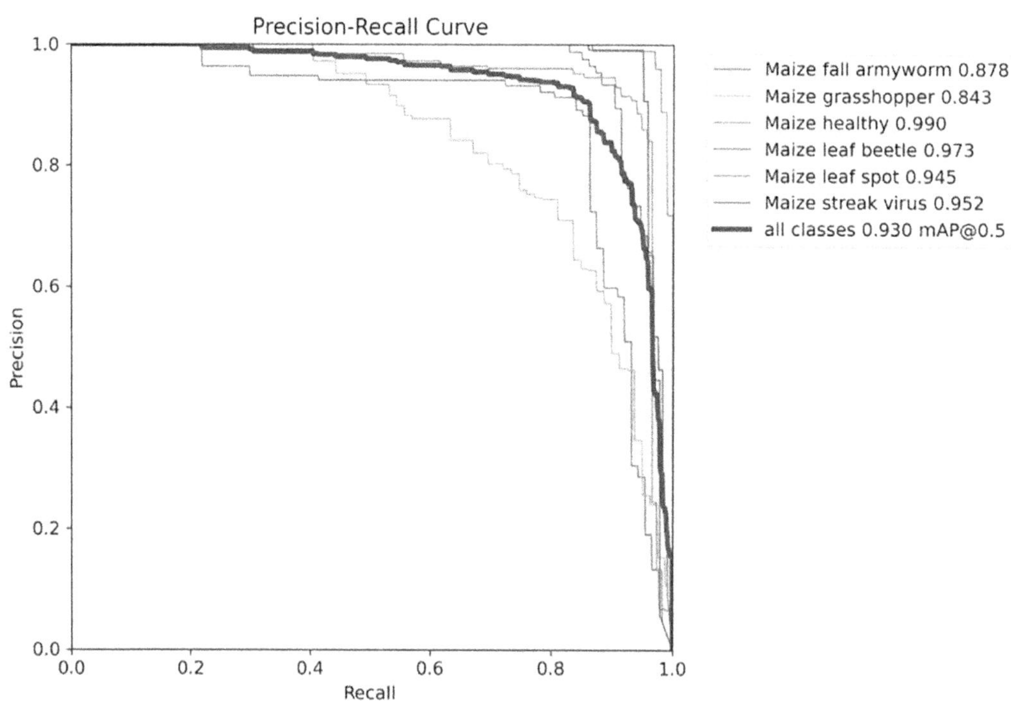

FIGURE 13.8 Precision-recall curve.

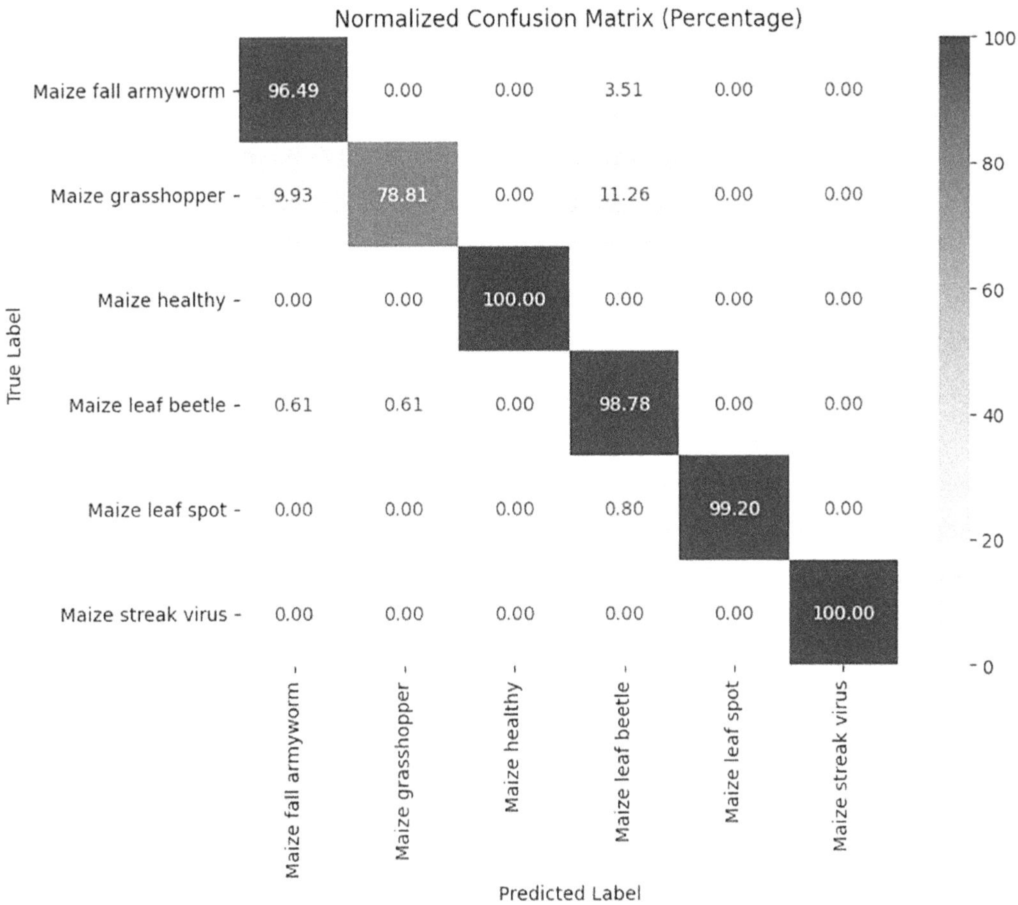

FIGURE 13.9 Hybrid model's confusion matrix.

13.5.5 Predictions

This section comprises a test of the prediction outcomes of VGG16, YOLOv11, and the Hybrid Model (YOLOv11 + VGG16) in order to determine their efficiency in identifying pests and diseases in maize. The detailed analysis identifies the degree to which the ability of each model to classify and detect various crop conditions is represented by evaluating the confidence scores and accuracy. VGG16 is mainly focused on classification, and thus it offers high confidence scores in different maize crop categories. On the other hand, YOLOv11 is a star in object detection. Accurately detecting and correctly locating pests and diseases with the use of bounding boxes.

The hybrid model combines the two abilities of detection and classification that make it possible to generate really accurate predictions.

Figure 13.10 presents the performance of VGG16 in classification by the identification of different maize crop conditions like pests, diseases, or healthy. There are high confidence scores in all the classes or categories that are used to predict maize fall armyworm, grasshopper, healthy maize, leaf beetle, leaf spot, and streak virus, which implies that the reliability of the model is at the peak performance of pest and disease detection in maize.

Table 13.1 lists down the various kinds of pests and diseases that attack maize and their corresponding prediction or confidence scores. The respective scores indicate each model's ability

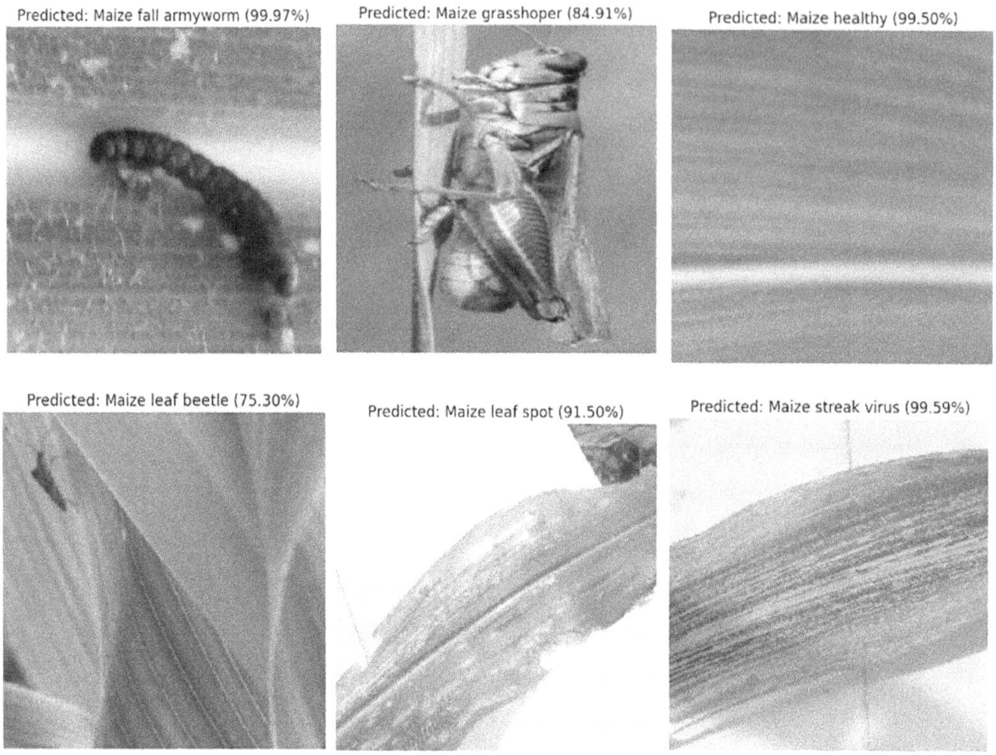

FIGURE 13.10 VGG16 models predictions.

TABLE 13.1
Prediction Confidence Scores for Maize Pests and Diseases

	Confidence Scores (%) of Pests and Diseases		
Predicted Class	**VGG16**	**YOLOv11**	**Hybrid Model (YOLOV11+ VGG16)**
Maize fall armyworm	99.97	96.50	99.91
Maize grasshopper	84.91	89.42	99.96
Maize healthy	99.50	98.82	98.59
Maize leaf beetle	75.30	93.55	99.73
Maize leaf spot	91.50	70.77	94.10
Maize streak virus	99.59	94.66	99.70

to classify different classes. It can be observed from Table 13.1 that the proposed hybrid model performs better compared to VGG16 and YOLOv11.

Figure 13.12 shows the detection results of the YOLO model for various maize pests and diseases. The YOLO model identifies Maize fall armyworms (96.50%), Maize grasshoppers (89.42%), Maize healthy (98.82%), Maize leaf beetles (93.55%), Maize leaf spots (70.77%), and Maize streak viruses (94.66%). Therefore, the scores are the prediction of the model to identify each category correctly.

As shown in Figure 13.12, the hybrid model distinguishes between different health statuses of the corn plant, e.g., pest attacks, diseases, and healthy leaves. Every prediction is made with a high

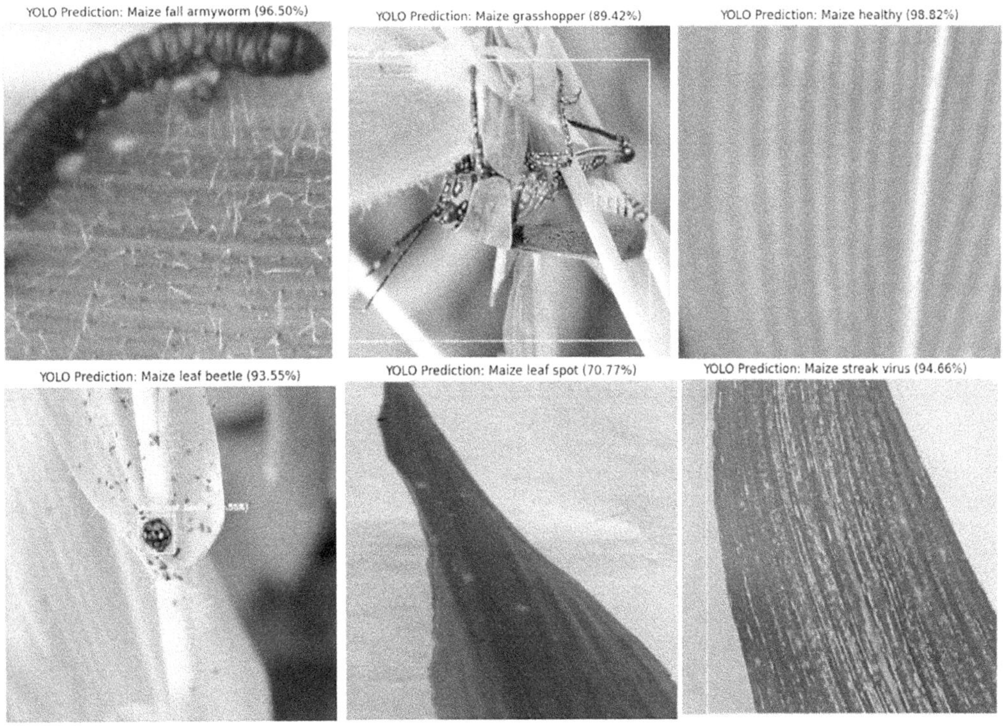

FIGURE 13.11 YOLOv11 models predictions.

FIGURE 13.12 Predictions of hybrid model.

TABLE 13.2
Comparison of Accuracy for VGG16, YOLOv11, and Hybrid Model

Model	Accuracy (%)
CNN	75.82
VGG16	90.63
YOLOv11	93.03
YOLOv5n (Coordinate Attention + Swin Transformer)	95.2
Random Forest	95
YOLO	94.5
YOLOv4	92.86
Hybrid (YOLOv11 + VGG16)	95.56

confidence score, and thus, the model's penultimate ability to tell one category from another is appreciated. Targeted maize fall armyworm, grasshopper, healthy, leaf beetle, leaf spot, and streak virus identification as well as their surrounding neighboring pest and diseases styles of detection are indications of the high accuracy of the model.

13.5.6 Comparison of the Results

Based on experimental outcomes, the hybrid model (YOLOv11 + VGG16) brought the highest accuracy at 95.56%, which outperformed CNN (75.82%), VGG16 (90.63%), and YOLOV11 (93.03%) as sown in Table 13.2. The YOLOv11 model displayed its excellent object detection properties by having a very high mAP@50 (0.930) and precision-recall balance that can achieve an efficient identification of maize pests and diseases.

On the other hand, VGG16 turned out to be the best when it came to classification accuracy, and this was why it was the most preferred one for fine-grained recognition of maize crop conditions. The hybrid model was smartly designed to take full advantages of each model's pro, and therefore high detection and classification performance with a little-to-no misclassifications.

Table 13.2 compares several models used for the detection of pests and diseases in maize crops. The standalone CNN model has a high misclassification rate (75.82%); even so, it does not have the object detection capability. For instance, VGG16, with its 90.63% accuracy, is a very good choice for classification but lacks object detection. The significant improvement in detection of YOLOv11 (i.e., 93.03%) is due to better localization and identification of pests and diseases, but it still faces challenges in classifying classes like maize leaf spot. The hybrid (YOLOv11 + VGG16) model reaches the best score (95.56%) that is obtained by employing YOLOv11 for object detection in combination with VGG16 for classification, thus minimizing the number of misclassifications. It achieves a better result than the other models, namely YOLOv5n (95.2%), RF (95%), and YOLO-based detection (94.5%). However, that needs a stronger machine. Still, in the area between detection and classification, it displays greatest potential for real-world agricultural applications. Nevertheless, the hybrid model calls for more computationally intensive hardware because of the combined architecture, and this could become a limitation in constrained computational environments

13.6 CONCLUSION

A hybrid model (YOLOv11 + VGG16) for pest and disease detection of maize was developed in this study, showing 95.56% accuracy, which is better than various existing models. While YOLOv11 was excellent in object detection (mAP@50: 0.930), VGG16 was strong in classification, which makes the hybrid model less prone to wrong classification. The hybrid approach was responsible for the improvement of accuracy by 5.12% against VGG16 and 2.53% against YOLOv11, thus confirming the fact that the integration of model multi-data improves detection performance. The focal point of this study is maize crop type, the hybrid technique is also capable of being applied to other crops and form a comprehensive pest and disease detection system.

REFERENCES

[1] Patil, Poonam L., Smita N. Chaudhri, and Nagama S. Kazzi. "Review on detection and prediction of crop disease and pest using machine learning." *Heritage Research Journal 71*, no. 12 (2023): 82–95.

[2] Dhaka, Vijaypal Singh, Nidhi Kundu, Geeta Rani, Ester Zumpano, and Eugenio Vocaturo. "Role of internet of things and deep learning techniques in plant disease detection and classification: A focused review." *Sensors* 23, no. 18 (2023): 7877.

[3] Nagar, Harshita, and R. S. Sharma. "A comprehensive survey on pest detection techniques using image processing." In *2020 4th International Conference on Intelligent Computing and Control Systems (ICICCS)*, pp. 43–48. IEEE, 2020.

[4] Dong, Shifeng, Jianming Du, Lin Jiao, Fenmei Wang, Kang Liu, Yue Teng, and Rujing Wang. "Automatic crop pest detection oriented multiscale feature fusion approach." *Insects* 13, no. 6 (2022): 554.

[5] Yang, Shuai, Ziyao Xing, Hengbin Wang, Xinrui Dong, Xiang Gao, Zhe Liu, Xiaodong Zhang, Shaoming Li, and Yuanyuan Zhao. "Maize-YOLO: A new high-precision and real-time method for maize pest detection." *Insects* 14, no. 3 (2023): 278.

[6] Echalar, Loyd, and Arnel C. Fajardo. "CNN's resnet, YOLO, and faster R-CNN architectures on the disease and pest classification of local agricultural vegetables towards sustainable production." *Available at SSRN 4487146* (2023).

[7] Ramesh, Shima, Ramachandra Hebbar, M. Niveditha, R. Pooja, N. Shashank, and P. V. Vinod. "Plant disease detection using machine learning." In *2018 International Conference on Design Innovations for 3Cs Compute Communicate Control (ICDI3C)*, pp. 41–45. IEEE, 2018.

[8] Masood, Momina, Marriam Nawaz, Tahira Nazir, Ali Javed, Reem Alkanhel, Hela Elmannai, Sami Dhahbi, and Sami Bourouis. "MaizeNet: A deep learning approach for effective recognition of maize plant leaf diseases." *IEEE Access* 11 (2023): 52862–52876.

[9] Li, Jing, and Shaobo Du. "A corn disease recognition and detection method based on convolutional neural network and YOLOv5." *Sciences* 61 (2024): 747–757.

[10] Singh, Shikha. "An improved plants disease prediction system based on Ann approach: A review."

[11] Singla, Annu, Ashima Nehra, Kamaldeep Joshi, Ajit Kumar, Narendra Tuteja, Rajeev K. Varshney, Sarvajeet Singh Gill, and Ritu Gill. "Exploration of machine learning approaches for automated crop disease detection." *Current Plant Biology 40* (2024): 1–13, 100382.

[12] Jackulin, C., and S. J. M. S. Murugavalli. "A comprehensive review on detection of plant disease using machine learning and deep learning approaches." *Measurement: Sensors* 24 (2022): 100441.

[13] Mondal, Biswajit, Megha Bhushan, Ishaan Dawar, Meghavi Rana, Arun Negi, and Shirshendu Layek. "Crop disease prediction using machine learning and deep learning: An exploratory study." In *2023 International Conference on Sustainable Computing and Smart Systems (ICSCSS)*, pp. 278–283. IEEE, 2023.

[14] Changji, Wen, Wang Qirui, and E. Hongrui. "Model for the recognition of large-scale multi-class diseases and pests." *Transactions of the Chinese Society of Agricultural Engineering,* 38, no. 8 (2022): 169–177.

[15] Senan, Norhalina, Muhammad Aamir, Rosziati Ibrahim, NSA M. Taujuddin, and WHN Wan Muda. "An efficient convolutional neural network for paddy leaf disease and pest classification." *International Journal of Advanced Computer Science and Applications* 11, no. 7 (2020).

[16] Lippi, Martina, Niccolò Bonucci, Renzo Fabrizio Carpio, Mario Contarini, Stefano Speranza, and Andrea Gasparri. "A YOLO-based pest detection system for precision agriculture." In *2021 29th Mediterranean Conference on Control and Automation (MED)*, pp. 342–347. IEEE, 2021.

[17] Domingues, Tiago, Tomás Brandão, and João C. Ferreira. "Machine learning for detection and prediction of crop diseases and pests: A comprehensive survey." *Agriculture* 12, no. 9 (2022): 1350.

[18] Francisco, Mauro, Fernando Ribeiro, Jose Metrolho, and Rogerio Dionisio. "Algorithms and models for automatic detection and classification of diseases and pests in agricultural crops: A systematic review." *Applied Sciences* 13, no. 8 (2023): 4720.

[19] Ayub, Umair, and Syed Atif Moqurrab. "Predicting crop diseases using data mining approaches: classification." In *2018 1st International Conference on Power, Energy and Smart Grid (ICPESG)*, pp. 1–6. IEEE, 2018.

[20] Murimi, Micheni Maurice. "Identification of maize leaf diseases using support vector machine and convolutional neural networks alexnet and resnet50." PhD diss., Karatina University, 2023.

[21] Ji, Zhengjie, Shudi Bao, Meng Chen, and Linjing Wei. "ICS-ResNet: A lightweight network for maize leaf disease classification." *Agronomy* 14, no. 7 (2024): 1587.

[22] Devi, R. S., V. R. Kumar, and P. Sivakumar. "EfficientNetV2 model for plant disease classification and pest recognition." *Computer Systems Science & Engineering* 45, no. 2 (2023): 2249.

[23] Al Mamun, Abdullah, David Ahmedt-Aristizabal, Miaohua Zhang, Md Ismail Hossen, Zeeshan Hayder, and Mohammad Awrangjeb. "Plant disease detection using self-supervised learning: A systematic review." *IEEE Access 12* (2024):171926–171943.

[24] Pang, Haitong, Yitao Zhang, Weiming Cai, Bin Li, and Ruiyin Song. "A real-time object detection model for orchard pests based on improved YOLOv4 algorithm." *Scientific Reports* 12, no. 1 (2022): 13557.

[25] Shoaib, Muhammad, Babar Shah, Shaker Ei-Sappagh, Akhtar Ali, Asad Ullah, Fayadh Alenezi, Tsanko Gechev, Tariq Hussain, and Farman Ali. "An advanced deep learning models-based plant disease detection: A review of recent research." *Frontiers in Plant Science* 14 (2023): 1158933.

[26] Ma, Li, Qiwen Yu, Helong Yu, and Jian Zhang. "Maize leaf disease identification based on YOLOv5n algorithm incorporating attention mechanism." *Agronomy* 13, no. 2 (2023): 521.

[27] Ariwa, Rosemary Ngozi, Joshua Babatunde Agbogun, Monday Jubrin Abdullahi, Temidayo Oluwatosin Omotehinwa, Bethel Ugochukwu Nnamani, Daniel Gabriel Zhetu, Umar Gambo Abubakar, and Kaduna North. "Comparative performance analysis of YOLOv8 model and traditional machine learning models for maize leaf disease detection."

[28] Satti, Hassan. "Enhancing maize disease detection: A comprehensive study on few-shot learning, object detection, and synthetic data generation using YOLOv5s-C3CBAM model." *International Journal of Agriculture and Sustainable Development* 5, no. 4 (2023): 186–197.

[29] Song, Yihong, Haoyan Zhang, Jiaqi Li, Ran Ye, Xincan Zhou, Bowen Dong, Dongchen Fan, and Lin Li. "High-accuracy maize disease detection based on attention generative adversarial network and few-shot learning." *Plants* 12, no. 17 (2023): 3105.

[30] Jain, Sonal, and Dharavath Ramesh. "AI based hybrid CNN-LSTM model for crop disease prediction: An ML advent for rice crop." In *2021 12th International Conference on Computing Communication and Networking Technologies (ICCCNT)*, pp. 1–7. IEEE, 2021.

[31] Karthik, R., Armaano Ajay, Akshaj Singh Bisht, T. Illakiya, and K. Suganthi. "A deep learning approach for crop disease and pest classification using swin transformer and dual-attention multi-scale fusion network." *IEEE Access 12* (2024): 152639–152655.

[32] Slimani, Hicham, Jamal El Mhamdi, and Abdelilah Jilbab. "Drone-assisted plant disease identification using artificial intelligence: A critical review." *International Journal of Computing and Digital Systems* 14, no. 1 (2023): 10433–10446.

[33] Rajeshram, V., B. Rithish, S. Karthikeyan, and S. Prathab. "Leaf diseases prediction pest detection and pesticides recommendation using deep learning techniques." In *2023 International Conference on Sustainable Computing and Data Communication Systems (ICSCDS)*, pp. 1633–1639. IEEE, 2023.

[34] Sharma, Vivek, Ashish Kumar Tripathi, and Himanshu Mittal. "Technological advancements in automated crop pest and disease detection: A review & ongoing research." In *2022 International Conference on Computing, Communication, Security and Intelligent Systems (IC3SIS)*, pp. 1–6. IEEE, 2022.

[35] Nishitha, Panchareddy, Tuniki Sarika, Soleti Navya, Subhashini Peneti, Dablikar Rajeshwari, and E. N. Vijayakumari. "Smart agriculture: CNN-based systems for early detection and diagnosis of plant diseases and pests." In *2024 Second International Conference on Intelligent Cyber Physical Systems and Internet of Things (ICoICI)*, pp. 1427–1431. IEEE, 2024.

[36] Paymode, Ananda S., and Vandana B. Malode. "Transfer learning for multi-crop leaf disease image classification using convolutional neural network VGG." *Artificial Intelligence in Agriculture* 6 (2022): 23–33.
[37] Lee, Yong-Suk, Maheshkumar Prakash Patil, Jeong Gyu Kim, Yong Bae Seo, Dong-Hyun Ahn, and Gun-Do Kim. "Hyperparameter optimization for tomato leaf disease recognition based on YOLOv11m." *Plants* 14, no. 5 (2025): 653.
[38] Rodríguez-Lira, Diana-Carmen, Diana-Margarita Córdova-Esparza, José M. Álvarez-Alvarado, Julio-Alejandro Romero-González, Juan Terven, and Juvenal Rodríguez-Reséndiz. "Comparative analysis of YOLO models for bean leaf disease detection in natural environments." *AgriEngineering* 6, no. 4 (2024): 4585–4603.

LIST OF ABBREVIATIONS

Abbreviation	Full Form
AI	Artificial Intelligence
ANN	Artificial Neural Networks
CNN	Convolutional Neural Network
DL	Deep Learning
FSL	Few-Shot Learning
GAN	Generative Adversarial Network
IoT	Internet of Things
KNN	k-Nearest Neighbors
LSTM	Long Short-Term Memory
ML	Machine Learning
RF	Random Forest
RNN	Recurrent Neural Network
SVM	Support Vector Machine
UAV	Unmanned Aerial Vehicle
VGG16	Visual Geometry Group 16-layer network
YOLO	You Only Look Once
IOU	Intersection Of Union

Index

Note: Page numbers in *italics* indicate figures and in **bold** indicate tables on the corresponding pages.

For Product Safety Concerns and Information please contact our EU
representative GPSR@taylorandfrancis.com
Taylor & Francis Verlag GmbH, Kaufingerstraße 24, 80331 München, Germany

www.ingramcontent.com/pod-product-compliance
Ingram Content Group UK Ltd.
Pitfield, Milton Keynes, MK11 3LW, UK
UKHW052220190926
475702UK00006B/1131
9781041008361